TRANSTHEORETIC FOUNDATIONS OF MATHEMATICS

(GENERAL SUMMARY OF RESULTS)

H. A. POGORZELSKI AND W. J. RYAN

SERIES I : NATURAL NUMBERS

VOLUME IB : ARITHMETICS

PREFACE

INTRODUCTION

Chapter 4
SEMIOLOGICAL ARITHMETICS

RESEARCH INSTITUTE FOR MATHEMATICS
383 COLLEGE AVENUE
ORONO, MAINE 04473

RIM Monographs in Mathematics

H. Pogorzelski
W. Ryan
W. Snyder
Editors

First Printing 1997.

ISBN 0-9643023-4-9 Research Institute for Mathematics

C O N T E N T S

VOLUME IA : FOUNDATIONS

VOLUME IB : ARITHMETICS

VOLUME IC : GOLDBACH CONJECTURE

PREFACE

When the series entitled Foundations of Semiological Theory of Numbers ([1]) (cf. Bibliography) was undertaken several dozens of years ago, we had in mind completing the whole series of some ten or so volumes within a decade. We considered and still consider the above-mentioned volumes and this new series to be but preliminary rough presentations of our Goldbach program, and only after their completion do we intend to polish the series down to several formal volumes. Alas, we have managed to publish only three of the projected ten volumes and this General Summary thus far. The unfortunate deprivation we have persistently encountered throughout the years in the way of research support, financial or otherwise, has taken a nontrivial toll on us and our work. Perhaps this was due in part to our failure to convincingly publicize some sort of preview of the whole panorama of our seemingly radical Goldbach Program. This General Summary is our requital to such an excuse.

In places where we took leave of the strict format of a General Summary, for the sake of brevity and time, due to the technical prerequisites involved, we were obliged to follow a somewhat uneven format in these volumes, citing in some cases nontrivial theorems with only hints of a proof and other less nontrivial cases with proofs. It should be noted that so-called transtheoretic proofs of results in our so-called trans-

---

([1]) This series has been terminated. It will be completely revised and incorporated in Series I: Natural Numbers.

theories are quite naturally far less technically involved than the proofs of such results would be if they could be carried out in so-called formal developments of mathematics. It should be noted that all proofs of a formal development of mathematics aligned in a transtheory are also proofs of the transtheory, but not conversely. At any rate, the formal details and much of the machinery already introduced earlier in these series but not mentioned in these volumes will be put into action in the future volumes of _RIM Series I, Natural Numbers_, and _RIM Series II, Irrational Numbers_.

INTRODUCTION

This summary, as in the case of the summaries of results of Bourbaki (see, for example, Bourbaki [1968, 1971]), is an informal description in the Riemann style [1867], rather than a formalization, of the results we plan to formalize in our series of projected research monographs, containing our transtheoretic proof [1] of the Goldbach Conjecture in a certain so-called Generalized Bourbaki-Church Transtheory, denoted as

$$\mathbb{T}_{GBC}: \mathcal{M}_{BS} \rightarrowtail \mathcal{F}_{BS},$$

the notation explained later. The preceding transtheory is an extension of the Bourbaki-Church Transtheory, denoted as

$$\mathbb{T}_{BC}: \mathcal{M}_{BS} \rightarrowtail \mathcal{F}_{BS},$$

which is an extension of the Bourbaki Transtheory, denoted as

$$\mathbb{T}_{B}: \mathcal{M}_{B} \rightarrowtail \mathcal{F}_{B},$$

where $\mathcal{M}_{BS}$ denotes a certain so-called Semiological Bourbaki Metaconceptual Theory, which is an extension of the Bourbaki Metaconceptual Theory $\mathcal{M}_{B}$, and $\mathcal{F}_{BS}$ denotes what we term the Semiological Bourbaki Formal Development of Mathematics, as depicted in the volumes by Pogorzelski and Ryan [1982-1988],

---

[1] Transtheoretic proof theory as assumed in this General Summary is not to be confused with transfinite proof theory of Gentzen [1936]. Our transtheoretic proof theory for General Recursive Arithmetic does not involve transfinite induction, especially over ordinals up to any ε-number, other than ordinary induction.

which is an extension of the Bourbaki Formal Development $\mathcal{F}_B$ as depicted in the work of the Bourbaki School (see §§1.1 and 1.3).

Because of the technicalities involved with $\mathbb{T}_{GBC}$, which will be introduced in Chapter 6 of Volume Ic, let's begin with the following motivation of the fundamental transtheory $\mathbb{T}_{BC}$. To describe Transtheory

$$\mathbb{T}_{BC}: \mathcal{M}_{BS} \rightarrowtail \mathcal{F}_{BS},$$

we offer the following motivation: Transtheory $\mathbb{T}_{BC}$ is a theoretical development of ideas surrounding Church's Thesis in that this thesis is only one of the axioms of $\mathbb{T}_{BC}$, which aligns objects of $\mathcal{M}_{BS}$ with objects of $\mathcal{F}_{BS}$ to be described in Chapter 3. We note that Church's Thesis, as well as certain well-formed formulas of $\mathbb{T}_{BC}$, are so-called alignments, denoted as $\mathfrak{m} \rightarrowtail \mathfrak{f}$, where the alignment sign $\rightarrowtail$ aligns metaobject $\mathfrak{m}$ of $\mathcal{M}_{BS}$ with formal object $\mathfrak{f}$ of $\mathcal{F}_{BS}$. Loosely speaking, an alignment of a transtheory is essentially a claim to the effect that the best possible candidate that can be semantically aligned with an essentially intuitively vague metaconcept $\mathfrak{m}$ of $\mathcal{M}$ is the well-defined formal object $\mathfrak{f}$ of $\mathcal{F}$. For example, Church's Thesis asserts the claim that the best candidate at explicating the metaconcept effective computability is either the well-defined formal concept partial recursiveness or general recursiveness. We note, if a different formal concept should be claimed as the best candidate for a transtheoretic alignment with effective computability, then, of course, in such a case we would be in effect proposing an entirely different transtheory from $\mathbb{T}_{BC}$. Interestingly, both Th. Skolem and D. Hilbert at one time believed that the best candidate for an alignment with effective computability should be <u>primitive recursiveness</u>.

Incidentally, the well-known unsolvability result concerning Hilbert's Tenth Problem (cf. M. Davis [1973]) is also not a result of $\mathcal{F}_B$ or $\mathcal{F}_{BS}$, but a result of the Bourbaki-Church Transtheory $\mathbb{T}_{BC}$. In fact, all results based on Church's Thesis are transtheoretic results. In turn, we note that the well-known Constructability Axiom V = L of Gödel [1940] is obviously an axiom of a transtheory expressible more correctly as the transtheoretic alignment $V \rightarrowtail L$. Cohen's [1963-1964] proof of the independence of $V \rightarrowtail L$ from the Zermelo-Fraenkel (ZF) axioms is an interesting clue to the possibility that all purely transtheoretic theorems and axioms of $\mathbb{T}_{BC}$ involving an alignment $\rightarrowtail$ are independent from ZF and its equivalencies and also from $\mathcal{F}_{BS}$.

We should note that clues for our resolution of the Goldbach Conjecture in $\mathbb{T}_{GBC}$ have been waiting in the literature for some time now (Pogorzelski [1969], [1970], [1974], [1976], [1977], [1977a] and Ryan [1974], [1976], [1978], [1979]). From the foundational point of view, it should be noted that the Goldbach Conjecture is a purely transtheoretic statement.

Incidentally, considering that the so-called Taniyama Conjecture is really only partially out of its incubator stage, Fermat's Last Theorem may be transtheoretic as well, which, semiologically speaking, essentially asserts that certain rather pathological structural properties of semisemiological spaces are not applicable to n-dimensional commutative semiological spaces for all $n > 2$ (cf. Pogorzelski and Ryan **[1982**, Chapter I, §3, no. 1, Definition 1, and Chapter I, §4, no. 1, Proposition 12, and §4, no. 4, Example 2]), involving not only the nonpower extension of the set of prime numbers (excluding the prime numbers themselves) and addition but all the rest of the infinite family of infinite sets $P^{(1)}$, $P^{(2)}$, ... of Hilbert-Ackermann numbers (cf. Pogorzelski [**1969-1970**] and §1.6, footnote 4) and addition.

As will be shown in this General Summary, the assumptions of the Goldbach Theorem of Pogorzelski [**1977a**] turn out to be theorems, inference rules, or axioms of Transtheory $\mathbb{T}_{GBC}$, and our transtheoretic proof of the Goldbach Conjecture in $\mathbb{T}_{GBC}$ involves relating the Goldbach Conjecture with Gödel's Second Incompleteness Theorem with respect to Equational General Recursive Arithmetic (see Ryan [**1976, 1978**]). Prime Number Theory enters the picture when we show in $\mathbb{T}_{GBC}$ that our arithmetics generate sums of odd prime numbers, based on showing that the set of all odd primes can be endowed with, contrary to the opinion of Gauss [**1801**], the structure of an unrooted infinite-dimensional noncommutative infinitely-rooted demisemiological prespace of a certain partition-theoretic number-model of the abstract infinite-dimensional commutative semiological space (cf. Pogorzelski and Ryan [**1982**]). Whereas D. Hilbert (Hilbert and Bernays [**1968-1970**]) attempted to prove the consistency of mathematics by essentially projecting the whole architecture of mathematics onto the metamathematical framework of a Primitive Recursive Arithmetic, we more modestly but analogously in dealing with our result essentially project the vast architecture of the Goldbach Problem onto the metamathematical framework of the even more general Equational General Recursive Arithmetic.

As we shall see, any formal development $\mathcal{F}$ based on the Bourbaki or ZF Theory of Sets without a subsuming transtheory $\mathbb{T}:\mathcal{M} \rightarrowtail \mathcal{F}$ constitutes an unfinished Theory-Holon (cf. §1.1), as already noted above. In consequence, considering that the natural habitat of any such $\mathcal{F}$ is $\mathbb{T}$, where $\mathbb{T}$-Provability is far stronger and more appropriate than $\mathcal{F}$-Provability, it would appear to be foundationally quite obtuse to insist on the existence of $\mathcal{F}$-proofs in $\mathcal{F}$ of $\mathbb{T}$-theorems as being a necessary prerequisite before accepting the mathematical validity of transtheoretic $\mathbb{T}$-proofs (i.e., proofs of $\mathbb{T}$ involving alignments) of such $\mathbb{T}$-theorems.

Lastly, we give an informal clue as to how transtheory $\mathbb{T}:\mathcal{M} \rightarrowtail \mathcal{F}$ locates with respect to the classical formalizations of foundations of mathematics. Let LG denote Logic, AR denote a natural-integer Arithmetic, and ST denote a Set Theory. Let LG-Ext [MAT-Ext] denote a theory or theories that are developed by extending an underlying Logic [mathematical theory or theories] by means of additional logical [mathematical] concepts. Let $\mathbb{T}$-Ext denote a transtheoretic extension to be presented later. We now diagrammatically summarize:

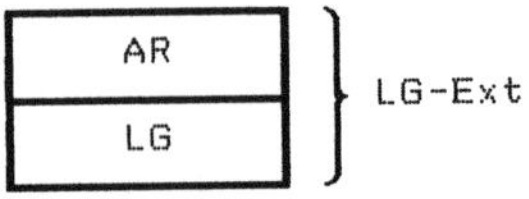

Frege [1884, 1893-1903]-Russell [1910-1913]

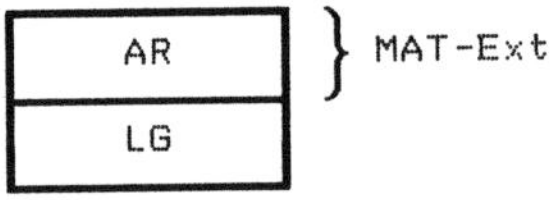

Dedekind [1888-1995]-Peano [1889]

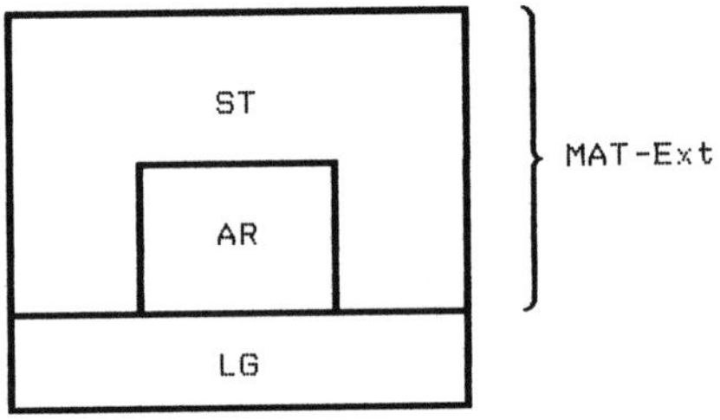

Dedekind [1889-1995]-Zermelo [1908]

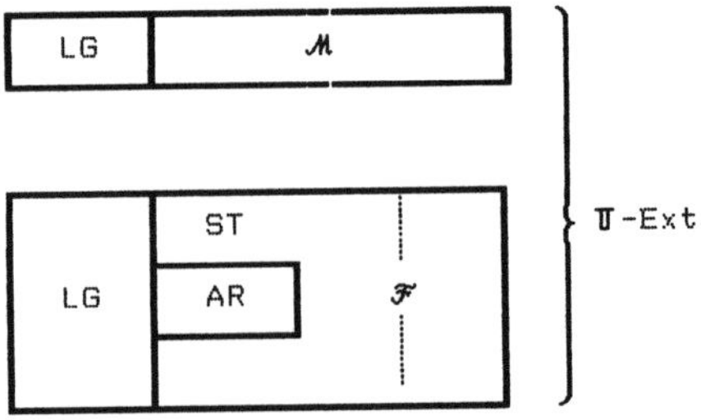

Transtheoretic Foundations

However gallant the efforts, the Frege-Russell foundations eventually foundered. The Dedekind-Peano foundations were absorbed in the Dedekind-Zermelo foundations. The Transtheoretic foundations essentially fill in the gaps left in the Dedekind-Zermelo foundations [2]. However, it should be noted that, unlike as in classical formalizations of foundations of mathematics, where only Logic, Set Theory and Arithmetic are subsumed, in Transtheoretic Foundations of Mathematics we further extend the sovereignty of the foundations of mathematics to include, on the one hand, General Topology and Theory of Topological Manifolds (cf. Bourbaki [**1939-present**]), and, on the other hand, General Semiology and Theory of Semiological Manifolds (cf. Pogorzelski and Ryan [**1982-1988**]).

Finally, we emphasize the following. Since we will make extensive use of the rubrics SEMIOLOGICAL and TOPOLOGICAL with respect to MATHEMATICS, THEORIES, STRUCTURES, SPACES and the like, we are obliged at this early stage to articulate their meanings within our transtheoretic format in order to avoid unnecessary confusion with the same rubrics as presently understood in the mathematical literature. Firstly, SEMIOLOGICAL and TOPOLOGICAL objects of any sort are assumed always to be part of Formal Development $\mathscr{F}_{\mathbf{BS}}$ of Mathematics, particularly

---

[2] This overdue recognition of Dedekind's pioneering contributions to set-theoretic foundations of mathematics (not to be confused with Cantor's transfinite set theory, as so-called historians do) is supported by Dedekind's classic [**1888**], becoming particularly evident when read in modern notation and terminology [**1995**].

when they are formally defined in $\mathcal{F}_{BS}$. From the point-of-view of a Transtheory $\mathbb{T}_{GBC}:\mathcal{M}_{BS} \rightarrowtail \mathcal{F}_{BS}$ or $\mathbb{T}_{BC}:\mathcal{M}_{BS} \rightarrowtail \mathcal{F}_{BS}$, the SEMIOLOGICAL and TOPOLOGICAL objects of $\mathcal{F}_{BS}$ will be assumed from two distinct perspectives, viz., formal and informal cases. Firstly, the formal case involves formally defined objects of $\mathcal{F}_{BS}$ that are formally aligned with instances of either the primitive metaconcept NEXTNESS or NEARNESS (cf. §2.1) of Metaconceptual Theory $\mathcal{M}_{BS}$ (cf. §2.1). Secondly, the informal case involves existing instances of NEXTNESS and NEARNESS that, for convenience and expediency, we can in obvious cases form informal alignments to SEMIOLOGICAL or TOPOLOGICAL objects that are as yet formally undefined in $\mathcal{F}_{BS}$. What is important here is that if the rubrics TOPOLOGICAL and SEMIOLOGICAL are used for whatever objects, formally defined or undefined in $\mathcal{F}_{BS}$, they can be assumed to be foundationally and semantically explicable by some instances of NEXTNESS or NEARNESS of $\mathcal{M}_{BS}$. For example, very generally asserting the matter, although we have not as yet formally defined the following theories, we may for the sake of providing a semantical crutch to understanding, assert the following. The SEMIOLOGICAL Theories of Natural Numbers, Integers and Rational Numbers in $\mathcal{F}_{BS}$ are semantically explicable only by some instances of NEXTNESS, while, in turn, the TOPOLOGICAL Theories of Real Numbers, Complex Numbers and the rest of the so-called Hypercomplex Numbers of $\mathcal{F}_{BS}$ are semantically explicable only by some instances of NEARNESS. Interestingly, the Theory of Prime Numbers and the Theory of Irrational Numbers are related, and both display similar erratic behaviors in their present linear habitats and an antipathy toward instances of both NEXTNESS and NEARNESS. Perhaps we have at last a signal to entirely new action-based primitive metaconcepts of $\mathcal{M}_{BS}$ other than NEXTNESS and NEARNESS. To sum, whenever the rubrics SEMIOLOGICAL and TOPOLOGICAL are used in whatever case, we assume an alignment with NEXTNESS and NEARNESS, respectively, and vice versa. These topics will be formally resumed in RIM Series II, Irrational Numbers.

CHAPTER 4. SEMIOLOGICAL ARITHMETICS ([1])

(TOPICS ON $\mathscr{F}$)

Some of the arithmetics in this chapter were already rather concisely introduced (Pogorzelski [1974]), which we include here for the record; however, not all such arithmetics are needed for Volume Ic.

§4.1. In this chapter we introduce monocursive, partial cursive, and cursive arithmetics, which we formalize as theories ([2]) in the extension $\mathscr{F}_{BS}$ of Bourbaki's Formal Development $\mathscr{F}_B$.

Recall the following from Pogorzelski and Ryan [1982,

([1]) This branch of mathematics was founded by Thoralf Skolem, who is still not fairly credited for his contributions to Set Theory, Logic, Foundations and Number Theory.

([2]) Not only these arithmetics but all the arithmetics introduced in Chapter 4 will be formalized as <u>theories</u> rather than as any of the structures cited in Chapter 1. We note that arithmetics are defined as theories involving terms, wffs, axioms, proofs, and theorems and hence are most appropriately Bourbaki theories (see Bourbaki [1968, Chapter I]). Bourbaki structures, on the other hand, are members of sets obtained from certain base sets by forming finitely many cartesian products and power sets of these sets (Bourbaki [1968, Chapter IV, §1, nos. 1-4]). Unfortunately, all too often we find the confusing circularity in the literature, when the underlying Peano space of an arithmetic is called an arithmetic without reference to either Dedekind's Iteration Theorem, axiomatic definitional schemes, or functional spaces. In turn, arithmetics are called systems, meaning structures. Semiological arithmetics are theories defined over semiological spaces. Recall (cf. §1.6), we extend the Bourbaki format, where a set endowed with a topological structure is called a <u>space</u>, to SEMIOLOGICAL MATHEMATICS by also calling a set endowed with a parasemiological structure a <u>space</u>, i.e., parasemiological space.

1985]:

Let S denote either a semiological space S (Pogorzelski and Ryan [**1982**, Chapter I, §4, no. 1, Definition 1]), a semicarrier space αSE, or a semicarrier space βES (Pogorzelski and Ryan [**1982**, Chapter I, §2, no. 4, Definition 2; see also Chapter I, §2, no. 5, Definition 1 and no. 6, Definition 1]), where E denotes the Peano space (Pogorzelski and Ryan [**1982**, Chapter I, §2, no. 3, Definition 8]).

Let $\mathbf{F}_{\mathbf{S}}^{***}$ denote either monocursive functional space $\mathbf{F}_{\mathbf{S}}^{\mathrm{mon}}$ (Pogorzelski and Ryan [**1985**, Chapter II, §2, no. 1, Definition 2]), cursive functional space $\mathbf{F}_{\mathbf{S}}^{\mathrm{cur}}$ (Pogorzelski and Ryan [**1985**, Chapter II, §3, no. 2, Definition 1]), or partial cursive functional space $\mathbf{F}_{\mathbf{S}}^{\mathrm{pcur}}$ (Pogorzelski and Ryan [**1985**, Chapter II, §3, no. 1, Definition 1]).

We thus see that $\mathbf{F}_{E}^{***}$ denotes either precursive functional space $\mathbf{F}_{E}^{\mathrm{pre}}$ (Pogorzelski and Ryan [**1985**, Chapter II, §4, no. 1, Definition 1]), recursive functional space $\mathbf{F}_{E}^{\mathrm{rec}}$ (Pogorzelski and Ryan [**1985**, Chapter II, §5, no. 3, Definition 1]), or partial recursive functional space $\mathbf{F}_{E}^{\mathrm{parec}}$ (Pogorzelski and Ryan [**1985**, Chapter II, §5, no. 1, Definition 1]), respectively. More generally, when in an argument we have *** = mon (resp. *** = cur; resp. *** = pcur), in those parts of the argument dealing with the Peano space E we have, by definition, *** = pre (resp. *** = rec; resp. *** = parec). (In other words, where $S = E$ we have mon = pre, cur = rec, and pcur = parec.)

Let *** $\in$ {mon, cur, pcur}. Since semicarriers αSE and

βES (see above references for these objects) are extensions of semiological spaces S, the functional spaces $\mathbf{F}^{***}_{\alpha SE}$ and $\mathbf{F}^{***}_{\beta ES}$ are extensions of functional spaces $\mathbf{F}^{***}_{S}$ defined similarly to the way in which we defined $\mathbf{F}^{***}_{S}$.

We recall (see footnote 1 to Definition 1 of §1.8 above) that we are extending the definition of "S-function" to sets S that are not finitistic. It is then clear that Proposition 13 and its corollaries on p. 0.4.6.21 of Pogorzelski and Ryan [1985] now hold for sets S that are not finitistic, so that

(*) <u>for any atomic set</u> S, <u>every nonempty</u> S-<u>function</u> f <u>is an</u> n-<u>place</u> S-<u>function for some unique</u> $n \geq 1$.

In turn, we recall that for any finitistic sets S and S', any partial function $\alpha: S \longrightarrow S'$, any $n \geq 1$, any n-place S-function f, and any set X of n-place S-functions, the notations

$$\alpha^{(n)}[f] \qquad \text{and} \qquad \alpha^{(n)}[X]$$

(written briefly, when no confusion will arise, as $\alpha[f]$ and $\alpha[X]$, respectively) are defined in Pogorzelski and Ryan [1982, Chapter 0, §1, no. 3]. We extend the definition of $\alpha[X]$ as follows:

DEFINITION 1. Let S be an atomic set (which includes the case where S is an atomic semiological space), let S' be any set, and let $\alpha: S \longrightarrow S'$ be a partial function. With set S being atomic, it follows by virtue of (*) above that for any set X of S-functions we are able to define the notation $\alpha[X]$ by:

$$\alpha[X] = \{\alpha^{(n)}[f] \mid f \in X \;\;\&\;\; n \geq 1 \;\;\&\;\; f \in \mathsf{F}_S^{(n)}\}.$$

§4.2. We motivate our higher-dimensional semiological arithmetics, to be introduced later, by introducing familiar one-dimensional arithmetics over the Peano space E. Our arithmetics will all be quantifier-free theories of $\mathcal{F}_{\mathbf{BS}}$, i.e., theories of $\mathcal{F}_{\mathbf{BS}}$ that do not involve either the universal or existential quantifiers, with these arithmetics obviously being subtheories of the familiar first-order theories of $\mathcal{F}_{\mathbf{BS}}$.

A sentential ***-arithmetic over E, denoted as SAR_E^{***}, which denotes either

$$\mathrm{SAR}_E^{pre}, \qquad \mathrm{SAR}_E^{rec}, \qquad \text{or} \qquad \mathrm{SAR}_E^{parec},$$

is a quantifier-free theory of $\mathcal{F}_{\mathbf{BS}}$ consisting of the following objects, formation rules, axioms, and inference rules:

(I) We assume the Peano space E.

(II) For $*** \in \{pre, rec, parec\}$, we assume the functional space F_E^{***} (we assume the theory of F_E^{***} (which includes first-order logic) to be in the metatheory of SAR_E^{***}).

(III) We assume the first-order primitive recursive arithmetic RA of Robbin [**1969**, pp. 72-110] to be in the metatheories of SAR_E^{rec} and SAR_E^{parec}.

(IV) We assume the definition of regular primitive recursive function given in Ryan [**1976**, Definition 1.1] to be in the metatheories of both SAR_E^{rec} and SAR_E^{parec}. In turn, for $n \geq 1$, we say that an $(n + 1)$-place primitive recursive function γ_E is partially regular if there exists a deduction **D** in RA of the wff

$$\forall x_1 \ldots \forall x_n [\exists y \exists z ([\gamma_E(x_1,\ldots,x_n,y) = 0 \ \& \ \gamma_E(x_1,\ldots,x_n,z) = 0] \Rightarrow [y = z])],$$

and we call **D** a partial-regularity deduction. In turn, we say that γ_E is strictly partially regular if it is partially regular but not regular. We then assume the definitions of both "partially regular primitive recursive function" and "strictly partially regular primitive recursive function" to be in the metatheory of $\mathrm{SAR}_E^{\mathrm{parec}}$.

(V) (1) The Primitive Recursive Formation Rules for $\mathrm{SAR}_E^{\mathrm{pre}}$ are formative criteria CF1, CF2, CF4, CF5, and CF6 of Bourbaki [1968, Chapter I, §1, no. 4].

(2) The Recursive [Partial Recursive] Formation Rules for $\mathrm{SAR}_E^{\mathrm{rec}}$ [$\mathrm{SAR}_E^{\mathrm{parec}}$] are formative criteria CF1, CF2, CF4, CF5, and CF6, and the following special case of CF3 [two following special cases of CF3] of Bourbaki [1968, Chapter I, §1, no. 4]:

(CF3′) For any $n \geq 1$ and any $(n + 1)$-place regular primitive recursive function $\gamma_E(x_1,\ldots,x_n,y)$ (see (IV) above), the formula

$$(*) \qquad \tau_y(\gamma_E(x_1,\ldots,x_n,y) = 0 \ \& \ [y \neq 0 \Rightarrow \prod_{i=0}^{y \dot{-} 1} \gamma_E(x_1,\ldots,x_n,i) \neq 0])$$

is a <u>term</u> of both $\mathrm{SAR}_E^{\mathrm{rec}}$ and $\mathrm{SAR}_E^{\mathrm{parec}}$, where $\dot{-}$ is restricted subtraction and Π is the product function (both defined in Robbin [**1969**, pp. 71, 91]). We denote (*) by the more familiar notation

$$\mu y(\gamma_E(x_1,\ldots,x_n,y) = 0).$$

(CF3″) For any $n \geq 1$, any $(n + 1)$-place partially regular primitive recursive function $\gamma_E(x_1,\ldots,x_n,y)$ (see (IV) above), and any constant terms $a_1, \ldots, a_n \in E$, if the relation

$$(\exists y)(\gamma_E(a_1,\ldots,a_n,y) = 0)$$

is a theorem of RA, then the formula

(**) $$\tau_y(\gamma_E(a_1,\ldots,a_n,y) = 0)$$

is a <u>term</u> of $\mathrm{SAR}_E^{\mathrm{parec}}$. We denote (**) by the more familiar notation

$$\iota y(\gamma_E(a_1,\ldots,a_n,y) = 0).$$

(VI) For *** $\in$ {pre,rec,parec}, the AXIOM SCHEMAS of SAR_E^{***} are Axiom Schemas S1, S2, S3, S4, and S6 of the sentential logic in Bourbaki [**1968**, Chapter I, §3, no. 1 and §5, no. 1].

(VII) For *** $\in$ {pre,rec,parec},

(i) all quantifier-free propositions of semiological number theory concerning Peano space E are axioms of SAR_E^{***} (see (I) above);

(ii) all defining equations of primitive recursive functions (including those of the projection and constant functions of E) are axioms of SAR_E^{***} (see (II) above).

(VIII) We have the following AXIOM:

AXIOM 1. $x = x$.

(IX) In the cases of $\mathrm{SAR}_E^{\mathrm{rec}}$ and $\mathrm{SAR}_E^{\mathrm{parec}}$ we have the following two AXIOM SCHEMAS:

AXIOM SCHEMA I. For any $n \geq 1$ and any $(n + 1)$-place regular primitive recursive function $\gamma_E(x_1,\ldots,x_n,y)$, the following relation is an axiom:

$$\gamma_E(x_1,\ldots,x_n,\mu y(\gamma_E(x_1,\ldots,x_n,y) = 0)) = 0.$$

AXIOM SCHEMA II. For any $n \geq 1$ and any $(n + 1)$-place regular primitive recursive function $\gamma_E(x_1,\ldots,x_n,y)$, the following relation is an axiom:

$$[\sigma_E(z) \dot{-} \mu y(\gamma_E(x_1,\ldots,x_n,y) = 0) = 0] \Rightarrow [\gamma_E(x_1,\ldots,x_n,z) \neq 0].$$

(X) In the case of $\mathrm{SAR}_E^{\mathrm{parec}}$ we have the following two additional AXIOM SCHEMAS:

AXIOM SCHEMA III. *For any $n \geq 1$, any $(n+1)$-place strictly partially regular primitive recursive function $\gamma_E(x_1,\ldots,x_n,y)$, and any constant terms $a_1, \ldots, a_n \in E$, if the relation*

$$(\exists y)(\gamma_E(a_1,\ldots,a_n,y) = 0)$$

is a theorem of RA, *then the following relation is an axiom of* $\mathrm{SAR}_E^{\mathrm{parec}}$:

$$\gamma_E(a_1,\ldots,a_n,\iota y(\gamma_E(a_1,\ldots,a_n,y) = 0)) = 0.$$

AXIOM SCHEMA IV. *For any $n \geq 1$, any $(n+1)$-place strictly partially regular primitive recursive function $\gamma_E(x_1,\ldots,x_n,y)$, and any constant terms $a_1, \ldots, a_n \in E$, if the relation*

$$(\exists y)(\gamma_E(a_1,\ldots,a_n,y) = 0)$$

is a theorem of RA, *then the following relation is an axiom of* $\mathrm{SAR}_E^{\mathrm{parec}}$:

$$[\sigma_E(z) \dot{-} \iota y(\gamma_E(a_1,\ldots,a_n,y) = 0) = 0] \Rightarrow [\gamma_E(a_1,\ldots,a_n,z) \neq 0].$$

(XI) We have the following INFERENCE RULES:

(1) The *criteria* C1 (*Modus Ponens*) and C3 (*Substitution of Terms for Variables*) of Bourbaki [1968, Chapter I, §2, nos. 2 and 3].

(2) The <u>Principle of Induction</u> (cf. C61 of Bourbaki [**1968**, Chapter III, §4, no. 3]). For any relation $R(x)$ of the arithmetic in question:

$$\frac{\begin{array}{c} R(0) \\ R(x) \Rightarrow R(\sigma_E(x)) \end{array}}{R(x)}$$

(3) No wff of $\mathscr{F}_{\mathbf{BS}}$ is a <u>theorem</u> of $\mathrm{SAR}_E^{\mathrm{pre}}$ unless it is derivable from the above axioms and axiom schemas using the above rules (except rules (CF3') and (CF3")). No wff of $\mathscr{F}_{\mathbf{BS}}$ is a theorem of $\mathrm{SAR}_E^{\mathrm{rec}}$ [$\mathrm{SAR}_E^{\mathrm{parec}}$] unless it is derivable from the axioms and axiom schemas of $\mathrm{SAR}_E^{\mathrm{rec}}$ [$\mathrm{SAR}_E^{\mathrm{parec}}$] by the rules of $\mathrm{SAR}_E^{\mathrm{rec}}$ [$\mathrm{SAR}_E^{\mathrm{parec}}$]. For *** ∈ {pre, rec, parec}, we let **Wff**(SAR_E^{***}) denote the set of <u>wffs</u> of SAR_E^{***} and let **Thm**(SAR_E^{***}) denote the set of <u>theorems</u> of SAR_E^{***}.

For *** ∈ {pre,rec,parec}, we have the following two propositions in SAR_E^{***}. They are proved by Bourbaki [**1968**, Chapter I, §5, no. 2, Theorems 2 and 3] using only his Theorem 1 of the same reference (which theorem we are assuming as an axiom of SAR_E^{***} (see (VIII) above)) and his axiom scheme S6 (which we are also assuming as an axiom of SAR_E^{***} (see (VI) above)):

PROPOSITION 1. $(x = y) \Rightarrow (y = x)$.

PROPOSITION 2. $((x = y)\ \&\ (y = z)) \Rightarrow (x = z)$.

Lastly, we have the following well-known result (which follows from the fact that the arithmetics in question can be Gödel-numbered):

PROPOSITION 3. For *** ∈ {pre,rec,parec}, we have:

(a) Card(**Wff**(SAR_E^{***})) is denumerably infinite;

(b) Card(**Thm**(SAR_E^{***})) is denumerably infinite.

§4.3. We denote as $\mathscr{S}_E^{pre}$ the sentential primitive recursive arithmetic developed in Ryan [**1974**] and extended in Ryan [**1979**]. One can prove in a straightforward manner that arithmetic $\mathscr{S}_E^{pre}$ is equivalent to the sentential primitive recursive arithmetic SAR_E^{pre} developed in the preceding section (see, in particular, Pogorzelski and Ryan [**1985**, Chapter II, §4, no. 1, Propositions 3, 4, and 5]).

In turn, we recall the sentential (general) recursive arithmetic $\mathscr{S}_E^{rec}$ defined in Ryan [**1976**, §2]. One proves in a straightforward manner that arithmetic $\mathscr{S}_E^{rec}$ is equivalent to sentential recursive arithmetic SAR_E^{rec} defined in the preceding section (note, in particular, that the pertinent properties of the μ-operator are given in Pogorzelski and Ryan [**1982**, Chapter 0, §0, no. 15, Proposition 1]).

Lastly, one obtains a sentential partial recursive arithmetic, which we denote as $\mathscr{S}_E^{parec}$, by making the obvious changes to $\mathscr{S}_E^{rec}$. One can then prove in a straightforward manner that arithmetic $\mathscr{S}_E^{parec}$ is equivalent to the sentential partial recursive arithmetic SAR_E^{parec} defined in the preceding section.

Then, letting $\mathscr{S}_E^{***}$ denote either $\mathscr{S}_E^{pre}$, $\mathscr{S}_E^{rec}$, or $\mathscr{S}_E^{parec}$, one can combine the preceding results into the following proposition:

PROPOSITION 1. The sentential arithmetics $\mathscr{S}_E^{***}$ are equivalent to the sentential arithmetics SAR_E^{***}.

§4.4. An equational ***-arithmetic over E, denoted as EAR_E^{***}, which denotes either

$$\mathrm{EAR}_E^{pre}, \qquad \mathrm{EAR}_E^{rec}, \qquad \text{or} \qquad \mathrm{EAR}_E^{parec},$$

is a logic-free theory of $\mathcal{F}_{\mathbf{BS}}$ consisting of the following objects, formation rules, axioms, and inference rules:

(I) We assume the Peano space E.

(II) For *** ∈ {pre,rec,parec}, we assume the functional space F_E^{***} (we assume the theory of F_E^{***} (which includes first-order logic) to be in the metatheory of EAR_E^{***}).

(III) We assume the first-order primitive recursive arithmetic RA of Robbin [1969, pp. 72-110] to be in the metatheories of EAR_E^{rec} and EAR_E^{parec}.

(IV) We assume the definition of "regular primitive recursive function" given in Ryan [1976, Definition 1.1] to be in the metatheories of EAR_E^{rec} and EAR_E^{parec}. In turn, we assume the definitions of both "partially regular primitive recursive function" and "strictly partially regular primitive recursive function" (given in (IV) of §4.2) to be in the metatheory of EAR_E^{parec}.

(V) We assume the Gödel numbering of RA given in Robbin [1969, pp. 100-102] to be in the metatheories of EAR_E^{rec} and EAR_E^{parec}.

(VI) We assume the primitive recursive function $\rho(n,i)$ that enumerates the Gödel numbers of $(n + 1)$-place _regular_ primitive recursive function constants (see Ryan [**1976**, Definition 1.7 and Proposition 1.8]) to be in the _metatheory_ of EAR_E^{rec}. In turn, we modify the definition of ρ in the obvious way to obtain a definition of a primitive recursive function $\rho'(n,i)$ that enumerates the Gödel numbers of $(n + 1)$-place _strictly partially regular_ primitive recursive function constants, and we assume ρ' to be in the _metatheory_ of EAR_E^{parec}.

(VII) The FORMATION RULES for formulas of EAR_E^{pre} are as follows:

(1) An individual variable is a _term_.

(2) A constant in E is a _term_.

(3) For any $n \geq 1$, any n-place primitive recursive function **f**, and any terms $\mathbf{t}_1$, $\mathbf{t}_2$, ..., $\mathbf{t}_n$, the expression $\mathbf{f}(\mathbf{t}_1,\mathbf{t}_2,\ldots,\mathbf{t}_n)$ is a _term_.

(4) For any terms $\mathbf{t}_1$ and $\mathbf{t}_2$, the expression $\mathbf{t}_1 = \mathbf{t}_2$ is a _wff_, with such a wff being called an _equation_.

(5) An expression is a term or wff (of EAR_E^{pre}) if and only if it is so by virtue of formation rules (1)-(4) above.

(VIII) The FORMATION RULES for formulas of EAR_E^{rec} are as follows:

(1′) An individual variable is a <u>term</u>.

(2′) A constant in E is a <u>term</u>.

(3′) For any $n \geq 1$, any n-place primitive recursive function **f**, and any terms $\mathbf{t}_1$, $\mathbf{t}_2$, ..., $\mathbf{t}_n$, the expression $\mathbf{f}(\mathbf{t}_1,\mathbf{t}_2,\ldots,\mathbf{t}_n)$ is a <u>term</u>.

(4′) For any $n \geq 1$, any $i \geq 0$, and any terms $\mathbf{t}_1$, $\mathbf{t}_2$, ..., $\mathbf{t}_n$, the expression $f_i^n(\mathbf{t}_1,\mathbf{t}_2,\ldots,\mathbf{t}_n)$ is a <u>term</u> ([1]).

(5′) For any terms $\mathbf{t}_1$ and $\mathbf{t}_2$, the expression $\mathbf{t}_1 = \mathbf{t}_2$ is a <u>wff</u>, with such a wff being called an <u>equation</u>.

(6′) An expression is a term or wff (of $\mathrm{EAR}_E^{\mathrm{rec}}$) if and only if it is so by virtue of formation rules (1′)-(5′) above.

(IX) The FORMATION RULES for formulas of $\mathrm{EAR}_E^{\mathrm{parec}}$ are as follows:

(1″) An individual variable is a <u>term</u>.

(2″) A constant in E is a <u>term</u>.

(3″) For any $n \geq 1$, any n-place primitive recursive function **f**, and any terms $\mathbf{t}_1$, $\mathbf{t}_2$, ..., $\mathbf{t}_n$, the expression $\mathbf{f}(\mathbf{t}_1,\mathbf{t}_2,\ldots,\mathbf{t}_n)$ is a <u>term</u>.

([1]) We use "f" instead of "F" for the general recursive function letter (cf. Ryan [1976, §3]) to avoid confusion with the notation for the infinite-dimensional commutative semiological space F.

(4″) For any $n \geq 1$, any $i \geq 0$, and any terms $\mathbf{t}_1$, $\mathbf{t}_2$, ..., $\mathbf{t}_n$, each of the expressions

$$f_i^n(\mathbf{t}_1,\mathbf{t}_2,\ldots,\mathbf{t}_n) \qquad \text{and} \qquad g_i^n(\mathbf{t}_1,\mathbf{t}_2,\ldots,\mathbf{t}_n)$$

is a <u>term</u>.

(5″) For any terms $\mathbf{t}_1$ and $\mathbf{t}_2$, the expression $\mathbf{t}_1 = \mathbf{t}_2$ is a <u>wff</u>, with such a wff being called an <u>equation</u>.

(6″) An expression is a term or wff (of $\mathrm{EAR}_E^{\mathrm{parec}}$) if and only if it is so by virtue of formation rules (1″)-(5″) above.

(X) For $*** \in \{\mathrm{pre},\mathrm{rec},\mathrm{parec}\}$, the AXIOMS and AXIOM SCHEMAS of EAR_E^{***} are the following:

(1) All quantifier-free propositions of semiological number theory concerning Peano space E and placed in the equation-calculus form (see Goodstein [**1957**, Chapter III]) are axioms of EAR_E^{***} (see (I) above).

(2) All defining equations of primitive recursive functions (including those of the projection and constant functions of E) are axioms of EAR_E^{***} (see (II) above).

(3) The following wff, which is a theorem of $\mathscr{F}_{\mathbf{BS}}$ (see Bourbaki [**1968**, p. 46, Theorem 1]), is an axiom of EAR_E^{***}:

AXIOM 1. $x = x$.

(4) In the cases of EAR_E^{rec} and EAR_E^{parec} we have the following two AXIOM SCHEMAS ([2]):

AXIOM SCHEMA I. For $n \geq 1$ and $i \geq 0$, if γ_E is the regular primitive recursive function constant with Gödel number $\rho(n,i)$, then the following equation is an axiom of EAR_E^{rec} and EAR_E^{parec}:

$$\gamma_E(x_1,\ldots,x_n,f_i^n(x_1,\ldots,x_n)) = 0.$$

AXIOM SCHEMA II. For $n \geq 1$ and $i \geq 0$, if γ_E is the regular primitive recursive function constant with Gödel number $\rho(n,i)$, then the following equation is an axiom of EAR_E^{rec} and EAR_E^{parec} ([3]):

$$(1 \dot{-} (\sigma_E(z) \dot{-} f_i^n(x_1,\ldots,x_n))) \times (1 \dot{-} \gamma_E(x_1,\ldots,x_n,z)) = 0.$$

(5) In the case of EAR_E^{parec} we have the following two

([2]) We point out that Axiom Schemas I and II are the equational equivalent to defining f_i^n by means of the familiar μ-operator (cf. Axiom Schemas I and II of §4.2 above). Namely, borrowing the notation of §4.2, we can say that

$$f_i^n(x_1,\ldots,x_n) = \mu y(\gamma_E(x_1,\ldots,x_n,y) = 0).$$

A similar remark applies to Axiom Schemas III and IV below.

([3]) Note, throughout the remainder of this volume we use the notations $\dot{-}$ and $\times$ to denote restricted subtraction and multiplication, respectively, defined in Robbin [1969, pp. 69, 71].

AXIOM SCHEMAS:

AXIOM SCHEMA III. For $n \geq 1$ and $i \geq 0$ and any constant terms $a_1, \ldots, a_n \in E$, if γ_E is the strictly partially regular primitive recursive function constant with Gödel number $\rho'(n, i)$ and if the relation

$$(\exists y)(\gamma_E(a_1, \ldots, a_n, y) = 0)$$

is a theorem of RA, then the following equation is an axiom of $\text{EAR}_E^{\text{parec}}$:

$$\gamma_E(a_1, \ldots, a_n, g_i^n(a_1, \ldots, a_n)) = 0.$$

AXIOM SCHEMA IV. For $n \geq 1$ and $i \geq 0$ and any constant terms $a_1, \ldots, a_n \in E$, if γ_E is the strictly partially regular primitive recursive function constant with Gödel number $\rho'(n, i)$ and if the relation

$$(\exists y)(\gamma_E(a_1, \ldots, a_n, y) = 0)$$

is a theorem of RA, then the following equation is an axiom of $\text{EAR}_E^{\text{parec}}$:

$$(1 \dot{-} (\sigma_E(z) \dot{-} g_i^n(a_1, \ldots, a_n))) \times (1 \dot{-} \gamma_E(a_1, \ldots, a_n, z)) = 0.$$

(XI) For *** $\in$ {pre, rec, parec}, the following are

INFERENCE RULES of EAR_E^{***}:

(1) The following two inference rules, which are derivable in $\mathcal{F}_{BS}$ (see Bourbaki [**1968**, p. 46, Theorems 2 and 3]):

$$(E_1) \qquad \frac{T = U}{U = T}$$

$$(E_2) \qquad \frac{\begin{array}{c} T = U \\ U = V \end{array}}{T = V}$$

(2) The following inference rule is derivable in a straightforward manner from Criterion C61 (<u>The Principle of Induction</u>) of Bourbaki [**1968**, p. 168]:

$$(IND) \qquad \frac{\begin{array}{c} \mathbf{f}(0) = \mathbf{g}(0) \\ \mathbf{f}(\sigma_E(x)) = H(x, \mathbf{f}(x)) \\ \mathbf{g}(\sigma_E(x)) = H(x, \mathbf{g}(x)) \end{array}}{\mathbf{f}(x) = \mathbf{g}(x)}$$

(3) EAR_E^{***} ($*** \in \{pre, rec, parec\}$) has the following two inference rules, the first of which is a special case of Criterion C3 of Bourbaki [**1968**, p. 26] and the second of which is Criterion C44 of Bourbaki [**1968**, p. 46]:

$$(S_1) \qquad \frac{\mathbf{f}(x) = \mathbf{g}(x)}{\mathbf{f}(T) = \mathbf{g}(T)}$$

$$(S_2) \qquad \frac{T = U}{\mathbf{f}(T) = \mathbf{f}(U)}$$

(4) In addition, in EAR_E^{rec} and EAR_E^{parec} we have the following Uniqueness Rule:

$$(U) \qquad \frac{\begin{array}{c} \mathbf{E}_1, \ldots, \mathbf{E}_k \\ \mathbf{E}_k \text{ is the formula } (0|x)\mathbf{E}_{k+1} \\ \mathbf{E}_{k+1} \vdash_{1,2} \mathbf{E}_{k+2}, \ldots, \mathbf{E}_{k+n-1}, (\sigma_E(x)|x)\mathbf{E}_{k+1} \end{array}}{\mathbf{E}_{k+1}}$$

where the $\mathbf{E}_i$ are equations, x is an individual variable, and $\vdash_{1,2}$ indicates a derivation by means of inference rules (S_1) and (S_2). The equations $\mathbf{E}_{k+1}, \ldots, \mathbf{E}_{k+n-1}, (\sigma_E(x)|x)\mathbf{E}_{k+1}$ are called conditional equations [4].

---

[4] In order to accommodate the conditional equations of the Uniqueness Rule (U) of our equational recursive arithmetic we amend our definition of a proof in a theory (see §1.2) so that in addition to the usual wffs in a proof

$$\mathfrak{w}_1, \mathfrak{w}_2, \ldots, \mathfrak{w}_n \quad (n \geq 1)$$

(where these wffs are all theorems), we allow one or more sequences of conditional wffs

$$\mathfrak{u}_1, \mathfrak{u}_2, \ldots, \mathfrak{u}_m \quad (m \geq 2),$$

where for each such sequence there exists a k, $1 \leq k < n$, such that, for some numerical variable x:

(XII) Let *** ∈ {pre, rec, parec}. Then no wff of $\mathcal{F}_{BS}$ is a theorem of EAR_E^{***} unless it is derivable from the axioms and axiom schemas of EAR_E^{***} using the inference rules of EAR_E^{***}. We let **Wff**(EAR_E^{***}) denote the set of <u>wffs</u> of EAR_E^{***} and let **Thm**(EAR_E^{***}) denote the set of <u>theorems</u> of EAR_E^{***}.

Lastly, we have the following well-known result (which follows from the fact that the arithmetics in question can be Gödel numbered):

PROPOSITION 1. <u>For</u> *** ∈ {pre,rec,parec}, <u>we</u> <u>have</u>:

(a) Card(**Wff**(EAR_E^{***})) <u>is</u> <u>denumerably</u> <u>infinite</u>;

(b) Card(**Thm**(EAR_E^{***})) <u>is</u> <u>denumerably</u> <u>infinite</u>.

(i) $\mathfrak{u}_1 = \mathfrak{w}_{k+1}$,

(ii) $\mathfrak{w}_k = (0|x)\mathfrak{w}_{k+1}$,

(iii) $\mathfrak{u}_m = (\sigma_E(x)|x)\mathfrak{w}_{k+1}$,

and where

(vi) $\mathfrak{u}_1$ (= $\mathfrak{w}_{k+1}$) is assumed as a hypothesis,

(v) for each i, $1 < i \leq m$, the wff $\mathfrak{u}_i$ is obtained from one of the wffs $\mathfrak{w}_1$, $\mathfrak{w}_2$, ..., $\mathfrak{w}_k$, $\mathfrak{u}_1$, $\mathfrak{u}_2$, ..., $\mathfrak{u}_{i-1}$ by one of the inference rules (S_1) and (S_2).

We extend this modification of the definition of a proof <u>mutatis</u> <u>mutandis</u> to the conditional equations in those of our further equational arithmetics that have a uniqueness rule.

§4.5. We denote as $\mathcal{G}_E^{pre}$ the <u>equational primitive recursive arithmetic</u> formalized in Goodstein [1957, pp. 104-112]. One can prove in a straightforward manner that arithmetic $\mathcal{G}_E^{pre}$ is equivalent to the equational primitive recursive arithmetic EAR_E^{pre} developed in the preceding section (note, in particular, that inference rule (IND) of EAR_E^{pre} is schema U_1 of $\mathcal{G}_E^{pre}$ (see Goodstein [1957, p. 105])).

In turn, we recall the <u>equational</u> (<u>general</u>) <u>recursive arithmetic</u> $\mathcal{G}_E^{rec}$ defined in Ryan [1976, §3]. One proves in a straightforward manner that arithmetic $\mathcal{G}_E^{rec}$ is equivalent to equational arithmetic EAR_E^{rec} defined in the preceding section (the usual properties of equality are obtained in $\mathcal{G}_E^{rec}$ as in Goodstein [1957, paragraph starting at bottom of p. 104]).

Lastly, one obtains an <u>equational partial recursive arithmetic</u>, which we denote as $\mathcal{G}_E^{parec}$, by making the obvious changes to $\mathcal{G}_E^{rec}$. One can then prove in a straightforward manner that $\mathcal{G}_E^{parec}$ is equivalent to the equational partial recursive arithmetic EAR_E^{parec} defined in the preceding section.

Then, letting $\mathcal{G}_E^{***}$ denote either $\mathcal{G}_E^{pre}$, $\mathcal{G}_E^{rec}$, or $\mathcal{G}_E^{parec}$, one can combine the preceding results into the following proposition:

PROPOSITION 1. <u>The equational arithmetics</u> $\mathcal{G}_E^{***}$ <u>are equivalent to the equational arithmetics</u> EAR_E^{***}.

Using the methods of Goodstein [1957, Chapter III], one easily defines the logical connectives $\neg$, $\vee$, &, $\Rightarrow$, $\Leftrightarrow$ in $\mathcal{G}_E^{pre}$, and then proves in a straightforward manner that the arithmetics $\mathcal{S}_E^{pre}$ and $\mathcal{E}_E^{pre}$ are equivalent.

In turn, it is proved in Ryan [**1976**, Main Theorem] that the arithmetics $\mathcal{S}_E^{rec}$ and $\mathcal{E}_E^{rec}$ are equivalent.

Lastly, using the methods of Ryan [**1976**], one proves that the arithmetics $\mathcal{S}_E^{parec}$ and $\mathcal{E}_E^{parec}$ are equivalent.

Then, combining the preceding results with Proposition 1 of §4.3 and Proposition 1 of the present section, one has the following proposition:

PROPOSITION 2. For *** $\in$ {pre, rec, parec}, the sentential arithmetics SAR_E^{***} are equivalent to the equational arithmetics EAR_E^{***}.

§4.6-4.7. For any semiological space S, a sentential monocursive arithmetic over S, denoted as $\mathrm{SAR}_S^{\mathrm{mon}}$, is a quantifier-free theory of $\mathcal{F}_{\mathbf{BS}}$ consisting of the following objects, formation rules, axioms, and inference rules:

(I) We assume the semiological space S in question.

(II) We assume the monocursive functional space $\mathbf{F}_S^{\mathrm{mon}}$ (see Pogorzelski and Ryan [**1985**, pp. II.2.1.7-8, Definition 2; see also pp. II.2.1.37-38, Proposition 13]).

(III) The Monocursive Formation Rules for $\mathrm{SAR}_S^{\mathrm{mon}}$ are formative criteria CF1, CF2, CF4, CF5, and CF6 of Bourbaki [**1968**, Chapter I, §1, no. 4].

(IV) The AXIOM SCHEMAS S1, S2, S3, S4, and S6 of the sentential logic in Bourbaki [**1968**, Chapter I, §3, no. 1, and §5, no. 1].

(V) The AXIOM:

AXIOM 1. $x = x$.

(VI) The following INFERENCE RULES:

(1) The criteria C1 (Modus Ponens) and C3 (Substitution of Terms for Variables) of Bourbaki [**1968**, Chapter I, §2, nos. 2 and 3].

(2) The Principle of Presemiological Induction (see Pogorzelski and Ryan [**1982**, p. I.1.1.21, Proposition 9]).

(3) No wff of $\mathcal{F}_{\mathbf{BS}}$ is a <u>theorem</u> of $\mathrm{SAR}_S^{\mathrm{mon}}$ unless it is derivable from the above axioms and axiom schemas using the above rules. We let **Wff**($\mathrm{SAR}_S^{\mathrm{mon}}$) denote the set of <u>wffs</u> of $\mathrm{SAR}_S^{\mathrm{mon}}$ and let **Thm**($\mathrm{SAR}_S^{\mathrm{mon}}$) denote the set of <u>theorems</u> of $\mathrm{SAR}_S^{\mathrm{mon}}$.

As in §4.2, we obtain the following two propositions:

PROPOSITION 1. $(x = y) \rightarrow (y = x)$.

PROPOSITION 2. $((x = y) \ \& \ (y = z)) \rightarrow (x = z)$.

§4.8. In this section we let S denote a semiological space with root set $R_S = \{\#_S\}$ and successor set

$$A_S = \{\sigma_1, \sigma_2, \sigma_3, \ldots\},$$

where A_S may be either finite (but nonempty) or denumerably infinite. (By virtue of S being a semiological space we have $S = \mathrm{cl}(S)$, $A_{\mathrm{cl}(S)} = A_S$, and $R_{\mathrm{cl}(S)} = R_S$ (Pogorzelski and Ryan [**1982**, p. I.4.1.3, Proposition 4]).)

Before formulating sentential cursive and partial cursive arithmetics, we introduce a monocursive analog of the first-order primitive recursive arithmetic RA of Robbin [**1969**, pp. 72-110]. The <u>first-order monocursive arithmetic over</u> S, which we denote as $\mathrm{RA}_S^{\mathrm{mon}}$, is obtained by retaining the axiom schemas 23.1-23.9 of Robbin [**1969**, p. 78] and replacing axiom schemas 23.10-23.17 of the same reference with the following axioms, axiom schemas, and inference rule:

AXIOM SCHEMA 1. <u>For all</u> $n \geq 1$ <u>and all</u> k, $1 \leq k \leq n$, <u>we have</u>

$$\mathrm{pr}_k^n(x_1,\ldots,x_n) = x_k.$$

AXIOM SCHEMA 2. <u>For all</u> $n \geq 1$ <u>and all</u> $\mathbf{b} \in S$, <u>we have</u>

$$\mathrm{c}_{\mathbf{b}}^n(x_1,\ldots,x_n) = \mathbf{b}.$$

AXIOM SCHEMA 3 (<u>Substitution Scheme</u>). <u>For all</u> $n \geq 1$ <u>and</u> $m \geq 1$, <u>all</u> $\mathrm{h} \in \mathrm{F}_S^{(n)}$ <u>and all</u> $\mathrm{g}_1, \ldots, \mathrm{g}_n \in \mathrm{F}_S^{(m)}$ <u>such that</u>

$$\mathrm{h}, \mathrm{g}_1, \ldots, \mathrm{g}_n \in \mathrm{F}_S^{\mathrm{mon}},$$

we have

$$\mathrm{Shg}_1 \ldots \mathrm{g}_n(x_1,\ldots,x_m) = \mathrm{h}(\mathrm{g}_1(x_1,\ldots,x_m),\ldots,\mathrm{g}_n(x_1,\ldots,x_m)).$$

AXIOM SCHEMA 4 (Monocursive Scheme). For all $n \geq 1$, all $\mathrm{g} \in \mathrm{F}_S^{(n)}$ such that $\mathrm{g} \in \mathrm{F}_S^{\mathrm{mon}}$, and all families

$$(\mathrm{h}_{\sigma(\#)})_{\sigma(\#) \in A_S^*} \quad [1],$$

such that for all $\sigma(\#) \in A_S^*$ we have

$$\mathrm{h}_{\sigma(\#)} \in \mathrm{F}_S^{(n+2)} \quad \text{and} \quad \mathrm{h}_{\sigma(\#)} \in \mathrm{F}_S^{\mathrm{mon}},$$

we have

$$\mathrm{Og}(\mathrm{h}_{\sigma(\#)})_{\sigma(\#) \in A_S^*}(x_1,\ldots,x_n,\#_S) = \mathrm{g}(x_1,\ldots,x_n)$$

$$\mathrm{Og}(\mathrm{h}_{\sigma(\#)})_{\sigma(\#) \in A_S^*}(x_1,\ldots,x_n,\sigma(y)) = \mathrm{h}_{\sigma(\#)}(x_1,\ldots,x_n,y, \mathrm{Og}(\mathrm{h}_{\sigma(\#)})_{\sigma(\#) \in A_S^*}(x_1,\ldots,x_n,y)) \quad (\sigma(\#) \in A_S^*).$$

In place of the axiom schemas 23.10-23.12 of Robbin [**1969**, p. 78] we have the axioms of the semiological space S in

---

[1] We have $A_S^* = \{\sigma(\#_S) \mid \sigma \in A_S \ \& \ \sigma \neq e_S\}$, where e_S is the identity successor function of S (in the event that space S is identified) (Pogorzelski and Ryan [**1982**, p. I.1.2.4, Definition 4]) (note, the conditions $\#_S \in R_S$ and $\sigma(\#_S) \in S$ of the cited definition are redundant here by virtue of S being a semiological space).

question (e.g., those axioms stating whether S is identified, zeroed, commutative, noncommutative, and the like); i.e., those axioms that determine the graph of space S.

As inference rules we have modus ponens and generalization (Robbin [**1969**, 11.6 and 11.7]) and the following (Pogorzelski and Ryan [**1982**, p. I.1.1.21, Proposition 9]):

PRESEMIOLOGICAL INDUCTION. For any relation $R(x)$, if

(i) $R(\#_S)$ holds, and

(ii) for all $x \in S$ and all $\sigma \in A_S$, if $R(x)$ holds, then $R(\sigma(x))$ holds,

then $R(x)$ holds for all $x \in S$.

We now have the following three definitions:

DEFINITION 1. We define the notion of n-place monocursive S-function inductively as follows:

(i) Each $\sigma \in A_S$ is a 1-place monocursive S-function.

(ii) For all $n \geq 1$ and all k, $1 \leq k \leq n$, pr_k^n is an n-place monocursive S-function.

(iii) For all $n \geq 1$ and all $\mathbf{b} \in S$, $c_{\mathbf{b}}^n$ is an n-place monocursive S-function.

(iv) For all $n \geq 1$ and all $m \geq 1$, if $\mathbf{h}$ is an n-place monocursive S-function and $\mathbf{g}_1, \ldots, \mathbf{g}_n$ are m-place monocursive S-functions, then

$$\mathbf{S}\mathbf{h}\mathbf{g}_1\ldots\mathbf{g}_n$$

is an m-place monocursive S-function.

(v) For all $n \geq 1$, if $\mathbf{g}$ is an n-place monocursive S-function and

$$(\mathbf{h}_{\sigma(\#)})_{\sigma(\#)} \in A_S^*$$

is a family such that for each $\sigma(\#) \in A_S^*$, $\mathbf{h}_{\sigma(\#)}$ is an $(n+2)$-place monocursive S-function, then

$$\mathbf{O}\mathbf{g}(\mathbf{h}_{\sigma(\#)})_{\sigma(\#)} \in A_S^*$$

is an $(n+1)$-place monocursive S-function.

DEFINITION 2. An $(n+1)$-place monocursive S-function γ_S $(n \geq 1)$ is said to be a <u>regular monocursive S-function</u> if and only if there exists a deduction **DED** in $\mathrm{RA}_S^{\mathrm{mon}}$ of the wff

$$\forall x_1\ldots\forall x_n\exists y\gamma_S(x_1,\ldots,x_n,y) = \#_S \ \ \& \ \ \forall x_1\ldots\forall x_n\forall y\forall z[(\gamma_S(x_1,\ldots,x_n,y) = \#_S \ \ \& \ \ \gamma_S(x_1,\ldots,x_n,z) = \#_S) \Rightarrow (y = z)],$$

in which case we call **DED** a <u>deduction of the regularity of</u> γ_S, an $(n+1)$-<u>place regularity deduction</u>, or briefly a <u>regularity deduction</u>.

In turn, it is easily seen that along the lines of Ryan [**1976**, §1] one can construct a Gödel numbering of arithmetic $\mathrm{RA}_S^{\mathrm{mon}}$. Then

(*) <u>one can define a primitive recursive function</u> $\rho_S(n, i)$ <u>that enumerates the Gödel numbers of</u> $(n+1)$-<u>place regular monocursive</u> S-<u>functions</u>.

DEFINITION 3. An $(n+1)$-place monocursive S-function γ_S $(n \geq 1)$ is said to be a <u>partially regular monocursive</u> S-<u>function</u> if and only if there exists a deduction **DED** in $\mathrm{RA}_S^{\mathrm{mon}}$ of the wff

$$\forall x_1 \ldots \forall x_n \forall y \forall z [(\gamma_S(x_1, \ldots, x_n, y) = \#_S \ \& \ \gamma_S(x_1, \ldots, x_n, z) = \#_S) \Rightarrow (y = z)],$$

in which case we call **DED** a <u>deduction of the partial regularity of</u> γ_S, an $(n+1)$-<u>place partial regularity deduction</u>, or briefly a <u>partial regularity deduction</u>.

An $(n + 1)$-place monocursive S-function is said to be a <u>strictly partially regular monocursive</u> S-<u>function</u> if it is partially regular but not regular.

In turn, one can use the Gödel numbering mentioned in Definition 2 above and, along the lines of Definition 2,

(*) <u>one can define a primitive recursive function</u> $\rho'_S(n, i)$ <u>that enumerates the Gödel numbers of</u> $(n+1)$-<u>place strictly partially regular monocursive</u> S-<u>functions</u>.

§4.9. In this section we let S denote a nonzeroed semiological space with root set $R_S = \{\#_S\}$ and successor set $A_S = \{\sigma_1, \sigma_2, \sigma_3, \ldots\}$, where A_S may be either finite (but nonempty) or denumerably infinite. We have $S = \mathrm{cl}(S)$, $A_{\mathrm{cl}(S)} = A_S$, and $R_{\mathrm{cl}(S)} = R_S$ (see first paragraph of §4.8).

A <u>sentential</u> *** <u>arithmetic</u> <u>over</u> S, denoted as SAR_S^{***}, which denotes either

$$\mathrm{SAR}_S^{\mathrm{cur}} \qquad \text{or} \qquad \mathrm{SAR}_S^{\mathrm{pcur}},$$

is a quantifier-free theory of $\mathcal{F}_{\mathrm{BS}}$ consisting of the following <u>objects</u>, <u>formation</u> <u>rules</u>, <u>axioms</u>, and <u>inference</u> <u>rules</u>:

(I) We assume the semiological space S given above.

(II) For *** ∈ {cur,pcur}, we assume the functional space $\mathbf{F}_S^{***}$ (we assume the <u>theory</u> of $\mathbf{F}_S^{***}$ (which includes first-order logic) to be in the <u>metatheory</u> of $\mathrm{SAR}_S^{\mathrm{cur}}$ and $\mathrm{SAR}_S^{\mathrm{pcur}}$).

(III) We assume the first-order monocursive arithmetic $\mathrm{RA}_S^{\mathrm{mon}}$ defined in §4.8 to be in the <u>metatheories</u> of both $\mathrm{SAR}_S^{\mathrm{cur}}$ and $\mathrm{SAR}_S^{\mathrm{pcur}}$.

(IV) We assume the definition of "regular monocursive S-function" given in §4.8 (Definitions 2 and 3) to be in the <u>metatheories</u> of $\mathrm{SAR}_S^{\mathrm{cur}}$ and $\mathrm{SAR}_S^{\mathrm{pcur}}$. In turn, we assume the definitions of both "partially regular monocursive S-function" and "strictly partially regular monocursive S-function" also given in §4.8 (Definitions 2 and 3) to be in the <u>metatheory</u> of $\mathrm{SAR}_S^{\mathrm{pcur}}$.

(V) The Cursive [Partial Cursive] Formation Rules for SAR_S^{cur} [SAR_S^{pcur}] are formative criteria CF1, CF2, CF4, CF5, and CF6, and the following special case [the two following special cases] of CF3 of Bourbaki [1968, Chapter I, §1, no. 4]:

(CF3*) For any $n \geq 1$ and any $(n+1)$-place regular monocursive S-function $\gamma_S(x_1,\ldots,x_n,y)$ (see §4.8 above), the formula

(*) $$\tau_y(\gamma_S(x_1,\ldots,x_n,y) = \#_S)$$

is a term of SAR_S^{cur} and SAR_S^{pcur}. We denote (*) by the more familiar notation

$$\iota y(\gamma_S(x_1,\ldots,x_n,y) = \#_S).$$

(CF3**) For any $n \geq 1$, any $(n + 1)$-place partially regular monocursive S-function $\gamma_S(x_1,\ldots,x_n,y)$ (see §4.8 above), and any constant terms $a_1, \ldots, a_n \in S$, if the relation

$$(\exists y)(\gamma_S(a_1,\ldots,a_n,y) = \#_S)$$

is a theorem of RA_S^{mon}, then the formula

(**) $$\tau_y(\gamma_S(a_1,\ldots,a_n,y) = \#_S)$$

is a term of SAR_S^{pcur}. We denote (**) by the more familiar notation

$$\iota y(\gamma_S(a_1,\ldots,a_n,y) = \#_S).$$

(VI) We assume the AXIOM SCHEMAS S1, S2, S3, S4, and S6 of the sentential logic in Bourbaki [**1968**, Chapter I, §3, no. 1, and §5, no. 1].

(VII) For *** ∈ {cur,pcur},

(i) all quantifier-free propositions of semiological number theory concerning semiological space S are axioms of SAR_S^{***} (see (I) above);

(ii) all defining equations of monocursive functions of S (including those of the projection and constant functions of S) are axioms of SAR_S^{***} (see (II) above).

(VIII) We assume the AXIOM:

AXIOM 1. $x = x$.

(IX) In the cases of $\mathrm{SAR}_S^{\mathrm{cur}}$ and $\mathrm{SAR}_S^{\mathrm{pcur}}$ we have the following AXIOM SCHEMAS:

AXIOM SCHEMA I. For any $n \geq 1$ and any $(n+1)$-place regular monocursive S-function $\gamma_S(x_1,\ldots,x_n,y)$, the following relation is an axiom:

$$\gamma_S(x_1,\ldots,x_n,\iota y(\gamma_S(x_1,\ldots,x_n,y) = \#_S)) = \#_S.$$

AXIOM SCHEMA II. For any $n \geq 1$ and any $(n+1)$-place regular monocursive S-function $\gamma_S(x_1,\ldots,x_n,y)$, the following relation is an axiom:

$$z \neq \iota y(\gamma_S(x_1,\ldots,x_n,y) = \#_S) \Rightarrow$$
$$\gamma_S(x_1,\ldots,x_n,z) \neq \#_S.$$

(X) In the case of $\mathrm{SAR}_S^{\mathrm{pcur}}$ we have the following two AXIOM SCHEMAS:

AXIOM SCHEMA III. For any $n \geq 1$, any $(n+1)$-place strictly partially regular monocursive S-function $\gamma_S(x_1,\ldots,x_n,y)$, and any constant terms $a_1, \ldots, a_n \in S$, if the relation

$$(\exists y)(\gamma_S(a_1,\ldots,a_n,y) = \#_S)$$

is a theorem of $\mathrm{RA}_S^{\mathrm{mon}}$, then the following relation is an axiom of $\mathrm{SAR}_S^{\mathrm{pcur}}$:

$$\gamma_S(a_1,\ldots,a_n,\iota y(\gamma_S(a_1,\ldots,a_n,y) = \#_S)) = \#_S.$$

AXIOM SCHEMA IV. For any $n \geq 1$, any $(n+1)$-place strictly partially regular monocursive S-function $\gamma_S(x_1,\ldots,x_n,y)$, and any constant terms $a_1, \ldots, a_n \in S$, if the relation

$$(\exists y)(\gamma_S(a_1,\ldots,a_n,y) = \#_S)$$

is a theorem of $\mathrm{RA}_S^{\mathrm{mon}}$, then the following relation is an axiom of $\mathrm{SAR}_S^{\mathrm{pcur}}$:

$$|z, \iota y(\gamma_S(a_1,\ldots,a_n,y) = \#_S)|_S \neq \#_S \Rightarrow$$
$$\gamma_S(a_1,\ldots,a_n,z) \neq \#_S$$

(where $|x,y|_S$ is the absolute difference function, which satisfies the conditions specified in Propositions 11 and 13 of §4.10 below).

(X) We have the following INFERENCE RULES:

(1) The criteria C1 (Modus Ponens) and C3 (Substitution of Terms for Variables) of Bourbaki [1968, Chapter I, §2, nos. 2 and 3].

(2) The Principle of Presemiological Induction (Pogorzelski and Ryan [1982, p. I.1.1.21, Proposition 9]):

PRESEMIOLOGICAL INDUCTION. For any relation $R(x)$, if

(i) $R(\#_S)$ holds, and

(ii) for all $x \in S$ and all $\sigma \in A_S$, if $R(x)$ holds, then $R(\sigma(x))$ holds,

then $R(x)$ holds for all $x \in S$.

(3) No wff of $\mathscr{F}_{\mathbf{BS}}$ is a theorem of $\mathrm{SAR}_S^{\mathrm{cur}}$ [$\mathrm{SAR}_S^{\mathrm{pcur}}$] unless it is derivable from the axioms and axiom schemas of $\mathrm{SAR}_S^{\mathrm{cur}}$ [$\mathrm{SAR}_S^{\mathrm{pcur}}$] by the rules of $\mathrm{SAR}_S^{\mathrm{cur}}$ [$\mathrm{SAR}_S^{\mathrm{pcur}}$].

As in §4.2, we obtain the following two propositions:

PROPOSITION 1. $(x = y) \Rightarrow (y = x)$.

PROPOSITION 2. $((x = y) \ \& \ (y = z)) \Rightarrow (x = z)$.

Lastly, recalling that the notations Wff(SAR_S^{mon}) and Thm(SAR_S^{mon}) were introduced in §4.6-4.7, we present the following definition:

DEFINITION 1. For *** ∈ {cur, pcur}, we let Wff(SAR_S^{***}) denote the set of <u>wffs</u> of SAR_S^{***} and let Thm(SAR_S^{***}) denote the set of <u>theorems</u> of SAR_S^{***}.

REMARK 1. We point out that if for all <u>zeroed</u> semiological spaces S (with root denoted as $\#_S$) there exist monocursive S-functions $|x,y|_S$ such that each of the relations

$$|x,y|_S = \#_S \qquad \text{and} \qquad x = y$$

is derivable from the other in both SAR_S^{mon} and EAR_S^{mon}, then Axiom Schema IV above would continue to hold if space S were zeroed, so that the preceding definition would be valid for <u>all</u> semiological spaces (and not only those that are nonzeroed).

Therefore, if Conjecture 1 of §4.10 above is true, then the preceding definition would not be valid if the semiological spaces S were allowed to be zeroed.

§4.10. In this section we introduce some preliminaries needed for the following sections.

Recall from Goodstein [1957, pp. 56-58] the following E-functions:

$$+,\ \dot{-},\ \times\ \in\ F_E^{pre} \qquad \text{and} \qquad |,|\ \in\ F_E^{pre}.$$

As treated in Goodstein [1957, pp. 50-58], we have the following, which relate wffs of EAR_E^{pre} and wffs of SAR_E^{pre}:

PROPOSITION 1.

(1) $|x,y| = \#_E$ iff $x = y$;

(2) $1 \dot{-} |x,y| = \#_E$ iff $x \neq y$;

(3) $|x,y| \times |z,w| = \#_E$ iff $[x = y \vee z = w]$;

(4) $|x,y| + |z,w| = \#_E$ iff $[x = y \ \&\ z = w]$;

(5) $(1 \dot{-} |x,y|) \times |z,w| = \#_E$ iff $[x = y \Rightarrow z = w]$;

(6) $(1 \dot{-} |x,y|) \times |z,w| +$
$(1 \dot{-} |z,w|) \times |x,y| = \#_E$ iff $[x = y \Leftrightarrow z = w]$.

Inasmuch as a semiological space is singly-rooted, for any semiological space S, with unique root $\#_S$, we are able to write the monocursive scheme (see Pogorzelski and Ryan [1985, pp. 0.4.3.1-2]) as:

$$(*)\quad \begin{aligned} f(X_n, \#_S) &= g(X_n) \\ f(X_n, \sigma(y)) &= h_{\sigma(\#_S)}(X_n, y, f(X_n, y)) \qquad (\sigma(\#_S) \in A_S^*), \end{aligned}$$

where $n \geq 1$, g is a previously-obtained n-place monocursive S-function, and

$$(h_{\sigma(\#_S)})_{\sigma(\#_S)} \in A_S^*$$

is a family of previously-obtained $(n + 2)$-place monocursive S-functions.

Then the following three propositions are proved in a straightforward manner (with each proof making use of presemiological induction (see Pogorzelski and Ryan [**1982**, p. I.1.1.21, Proposition 9])):

PROPOSITION 2. If S is a nonzeroed noncommutative semiological space, then the monocursive scheme (*) defines a total $(n + 1)$-place monocursive S-function.

PROPOSITION 3. If S is a commutative semiological space with root $\#_S$, then the monocursive scheme defines a total $(n + 1)$-place monocursive S-function if and only if for all $x_1, x_2, \ldots, x_n, y \in S$, all $\sigma(\#_S) \in A_S^*$, and all $\tau(\#_S) \in A_S^*$, the following relation holds:

$$h_{\sigma(\#_S)}(X_n, \tau(y), f(X_n, y)) = h_{\tau(\#_S)}(X_n, \sigma(y), f(X_n, y))$$

(cf. Vučković [**1959**, p. 306, paragraphs containing (2.3), (2.3'), and (2.3")]).

PROPOSITION 4. If S is a zeroed semiological space with root $\#_S$ and zeroed successor function ζ, then the monocursive scheme defines a total $(n + 1)$-place monocursive S-function if and only if for all $x_1, x_2, \ldots, x_n \in S$, the following condition holds:

$$h_{\zeta(\#_S)}(X_n, \#_S, g(X_n)) = g(X_n).$$

For the remainder of this section we let S be a semiological space with successor set

$$A_S = \{\sigma_1, \sigma_2, \sigma_3, \ldots\},$$

where A_S is permitted to be either finite (but nonempty) or denumerably infinite, and root set $R_S = \{\#_S\}$. We may, without loss of generality, assume that $\sigma_1(\#_S) \notin R_S$ (Pogorzelski and Ryan [1982, p. I.1.1.1, Axiom PS5]), in which case $\sigma_1(\#_S) \neq \#_S$. Lastly, we will let ζ denote the zeroed successor function of space S, in the event that space S is zeroed.

Now, we define five monocursive functions that yield the usual logical connectives $\neg$, $\vee$, &, $\Rightarrow$, and $\Leftrightarrow$ in EAR_S^{***}, the monocursive, cursive, and partial cursive equational arithmetics over S (in each case, one uses Propositions 2-4 above to prove that the following five definitions do, in fact, define total 1-place (in the case of $\overline{\mathrm{sg}}_S$) and 2-place monocursive S-functions):

(I) We define a 2-place monocursive S-function $\beta_S(x, y)$ by the equations:

$$\beta_S(x, \#_S) = \sigma_1(\#_S)$$

$$\beta_S(x, \sigma(y)) = \begin{cases} \#_S & \text{if } \sigma \neq \zeta \\ \beta_S(x, y) & \text{if } \sigma = \zeta, \end{cases}$$

and then we define a 1-place monocursive S-function $\overline{\mathrm{sg}}_S(x)$ (which yields <u>negation</u>) by the substitution scheme (see Pogorzelski and Ryan [**1985**, pp. 0.4.2.1-5]):

$$\overline{\mathrm{sg}}_S(x) = \beta_S(\mathrm{pr}_1^1(x), \mathrm{pr}_1^1(x)) \quad (= \beta_S(x, x)),$$

and we get the following:

PROPOSITION 5. <u>For any semiological space</u> S <u>with root</u> $\#_S$,

$$\overline{\mathrm{sg}}_S(x) = \#_S \iff x \neq \#_S \qquad (x \in S).$$

(II) We define a 2-place monocursive S-function $\mathbf{disj}_S(x, y)$ (which yields <u>disjunction</u>) by the monocursive scheme:

$$\mathbf{disj}_S(x, \#_S) = \#_S$$

$$\mathbf{disj}_S(x, \sigma(y)) = \begin{cases} x & \text{if } \sigma \neq \zeta \\ \mathbf{disj}_S(x, y) & \text{if } \sigma = \zeta, \end{cases}$$

and we get the following:

PROPOSITION 6. <u>For any semiological space</u> S <u>with root</u> $\#_S$,

$$\mathbf{disj}_S(x, y) = \#_S \iff [x = \#_S \vee y = \#_S] \qquad (x, y \in S).$$

(III) We define a 2-place monocursive S-function **conj**$_S(x,y)$ (which yields <u>conjunction</u>) by the monocursive scheme:

$$\mathbf{conj}_S(x,\#_S) = x$$

$$\mathbf{conj}_S(x,\sigma(y)) = \begin{cases} \sigma_1(\#_S) & \text{if } \sigma \neq \zeta \\ \mathbf{conj}_S(x,y) & \text{if } \sigma = \zeta, \end{cases}$$

and we get the following:

PROPOSITION 7. <u>For any semiological space</u> S <u>with root</u> $\#_S$,

$$\mathbf{conj}_S(x,y) = \#_S \leftrightarrow [x = \#_S \ \& \ y = \#_S] \quad (x,\ y \in S).$$

(IV) We define a 2-place monocursive S-function **imp**$_S(x,y)$ (which yields <u>implication</u>) by the monocursive scheme:

$$\mathbf{imp}_S(x,\#_S) = \#_S$$

$$\mathbf{imp}_S(x,\sigma(y)) = \begin{cases} \overline{\mathrm{sg}}_S(x) & \text{if } \sigma \neq \zeta \\ \mathbf{imp}_S(x,y) & \text{if } \sigma = \zeta, \end{cases}$$

and we get the following:

PROPOSITION 8. <u>For any semiological space</u> S <u>with root</u> $\#_S$,

$$\mathbf{imp}_S(x,y) = \#_S \leftrightarrow [x = \#_S \rightarrow y = \#_S] \quad (x,\ y \in S).$$

(V) We define a 2-place monocursive S-function **equiv**$_S(x,y)$ (which yields <u>equivalence</u>) by the monocursive scheme:

$$\mathrm{equiv}_S(x,\#_S) = x$$

$$\mathrm{equiv}_S(x,\sigma(y)) = \begin{cases} \overline{\mathrm{sg}}_S(x) & \text{if } \sigma \neq \zeta \\ \mathrm{equiv}_S(x,y) & \text{if } \sigma = \zeta, \end{cases}$$

and we get the following:

PROPOSITION 9. _For any semiological space_ S _with root_ $\#_S$,

$$\mathrm{equiv}_S(x,y) = \#_S \leftrightarrow [x = \#_S \leftrightarrow y = \#_S] \quad (x,\ y \in S).$$

We recall that Vučković [1962, §2] allows an alphabet of any cardinality in his word arithmetics, and hence these alphabets are permitted to be either finite or denumerably infinite ([1]). From this, along with the other results in Vučković [1962, §§1-3] and Proposition 2 above, we obtain the following:

PROPOSITION 10. _The set_ F_S^{mon} _of monocursive functions over any nonzeroed noncommutative semiological space_ S _can be considered to be precisely the set of primitive recursive functions of a Vučković primitive recursive arithmetic of words._

Moreover, a sentential [equational] monocursive arithmetic SAR_S^{mon} [EAR_S^{mon}] _over a nonzeroed noncommutative semiological space_ S _can be considered to be a sentential [equational] primitive recursive arithmetic of words of_

([1]) We note that Vučković's recursive arithmetics of words are monocursive arithmetics over magmatical spaces (cf. Pogorzelski and Ryan [1982, Chapter I, §4, no. 3]).

Vučković [1962, §3] (2).

We see (Vučković [1962, §3]) that in each of Vučković's primitive recursive arithmetics of words (denote an arbitrary one as V, say) there is a primitive recursive word function which we can denote as $|x,y|_V$ such that each of the relations

$$|x,y|_V = \#_V \qquad \text{and} \qquad x = y$$

is derivable from the other in the Vučković primitive recursive arithmetic V.

From this fact, along with the result

$$F_S^{mon} \leq F_S^{cur} \leq F_S^{pcur}$$

(Pogorzelski and Ryan [1985, pp. II.6.1.11-12, Proposition 2]), using Proposition 10 above, we obtain the following proposition:

PROPOSITION 11. For any nonzeroed noncommutative semiological space S, with root $\#_S$, there exists a monocursive function, which we will denote as $|x,y|_S$ such that each of the relations

$$|x,y|_S = \#_S \qquad \text{and} \qquad x = y$$

is derivable from the other in any sentential [equational] arithmetic SAR_S^{***} [EAR_S^{***}] over S (*** ∈ {mon, cur, pcur}).

---

(2) Although Vučković [1962] does not formalize his arithmetics, his proofs seem in the nature of proofs in an equational calculus. We note, however, that one could easily formulate Vučković's primitive recursive arithmetics of words as either sentential or equational calculi.

REMARK 1. We point out that Proposition 11, in the case where space S is _finite-dimensional_, can be proved using the results of Asser [1960].

We recall that Vučković [1959, §2] allows either a finite or a denumerably infinite number of successor functions in his partially ordered recursive arithmetics (which are _commutative_ arithmetics).

Then observing the other results of Vučković [1959], we obtain the following proposition:

PROPOSITION 12. _The set_ F_S^{mon} _of monocursive functions over any commutative semiological space_ S _can be considered to be precisely the set of primitive recursive functions of a Vučković partially ordered recursive arithmetic_.

Moreover, an equational monocursive arithmetic EAR_S^{mon} _over a commutative semiological space_ S _can be considered to be a partially ordered recursive arithmetic of Vučković_ [1959] (_which is a primitive recursive equational arithmetic_).

We recall that (see Vučković [1959, §§2-8]) in each of Vučković's partially ordered recursive arithmetics (denote an arbitrary one as V, say) there is a primitive recursive function, which we can denote as $|x,y|_V$, such that each of the relations

$$|x,y|_V = 0 \qquad \text{and} \qquad x = y$$

is derivable from the other in the Vučković partially ordered recursive arithmetic V.

From this fact, along with the result

$$F_S^{mon} \leq F_S^{cur} \leq F_S^{pcur}$$

(Pogorzelski and Ryan [**1985**, pp. II.6.1.11-12, Proposition 2]), using Proposition 12 above, we obtain the following proposition:

PROPOSITION 13. _For any commutative semiological space_ S, _with root_ $\#_S$, _there exists a monocursive function, which we will denote as_ $|x,y|_S$, _such that each of the relations_

$$|x,y|_S = \#_S \qquad \text{and} \qquad x = y$$

is derivable from the other in any equational arithmetic EAR_S^{***} ($*** \in$ {mon, cur, pcur}).

Lastly, we leave it as an open question whether, in the case of a _zeroed_ semiological space S, with root $\#_S$, there exists a monocursive function $|x,y|_S$ such that each of the relations

$$|x,y|_S = \#_S \qquad \text{and} \qquad x = y$$

is derivable from the other in any sentential [equational] arithmetic SAR_S^{***} [EAR_S^{***}] ($*** \in$ {mon, cur, pcur}).

Inasmuch as the condition

$$h_{\zeta(\#_S)}(X_n, \#_S, g(X_n)) = g(X_n)$$

given in Proposition 4 above seems not to hold for a great many functions $h_{\zeta(\#_S)}$, we make the following two conjectures:

CONJECTURE 1. For any zeroed semiological space S, with root $\#_S$, there does not exist a monocursive function $|x,y|_S$ such that each of the relations

$$|x,y|_S = \#_S \qquad \text{and} \qquad x = y$$

is derivable from the other in any sentential [equational] arithmetic SAR_S^{***} [EAR_S^{***}] ($*** \in$ {mon, cur, pcur}).

CONJECTURE 2. If the second equation of the monocursive scheme (see Pogorzelski and Ryan [**1985**, pp. 0.4.3.1-2], along with the fourth paragraph of the present subsection) were modified to read

$$f(X_n,\sigma(y)) = h_{\sigma(\#)}(X_n,y,f(X_n,y)) \qquad (\sigma(\#) \in A_S^*,\ \sigma(y) \notin R_S)$$

(that is, if the condition $\sigma(y) \notin R_S$ were added to the second equation, thus eliminating the "overlap" between the two equations in the monocursive scheme), then for any zeroed semiological space S, with root $\#_S$, there would exist a monocursive function $|x,y|_S$ such that each of the two relations

$$|x,y|_S = \#_S \qquad \text{and} \qquad x = y$$

is derivable from the other in any sentential [equational] arithmetic SAR_S^{***} [EAR_S^{***}] ($*** \in$ {mon, cur, pcur}).

(We note, moreover, that the monocursive scheme thus

modified would not change the content of the set $\mathbf{F}_S^{mon}$ of monocursive S-functions in the case where S is a nonzeroed semiological space.)

With Propositions 5, 8, 11, and 13, we have the following:

PROPOSITION 14. For any commutative or nonzeroed noncommutative semiological space S, with root $\#_S$, for any terms S, T, U, and V, the relation

$$\mathbf{imp}_S(\overline{sg}_S(|S,T|_S),\overline{sg}_S(|U,V|_S)) = \#_S$$

is equivalent to the relation

$$[S \neq T] \Rightarrow [U \neq V].$$

§4.11. In this section we let S denote either a commutative or a nonzeroed noncommutative semiological space with root set $R_S = \{\#_S\}$ and successor set

$$A_S = \{\sigma_1, \sigma_2, \sigma_3, \ldots\},$$

where A_S may be either finite (but nonempty) or denumerably infinite. (Thus σ_1 is a fixed successor function for which we may, without loss of generality, assume that $\sigma_1(\#_S) \neq \#_S$ (Pogorzelski and Ryan [1982, p. I.1.1.1, Axiom PS5]). We have $S = \mathrm{cl}(S)$, $A_{\mathrm{cl}(S)} = A_S$, and $R_{\mathrm{cl}(S)} = R_S$ (see first paragraph of §4.8).

An equational ***-arithmetic over S, denoted as EAR_S^{***}, which denotes either

$$\mathrm{EAR}_S^{\mathrm{mon}}, \qquad \mathrm{EAR}_S^{\mathrm{cur}}, \qquad \text{or} \qquad \mathrm{EAR}_S^{\mathrm{pcur}},$$

is a logic-free theory of $\mathscr{F}_{\mathbf{BS}}$ consisting of the following objects, formation rules, axioms, and inference rules:

(I) We assume the semiological space S given above.

(II) For *** $\in$ {mon, cur, pcur}, we assume the functional space F_S^{***} (we assume the theory of F_S^{***} (which includes first-order logic) to be in the metatheory of EAR_S^{***}).

(III) We assume the first-order monocursive arithmetic $\mathrm{RA}_S^{\mathrm{mon}}$ defined in §4.8 to be in the metatheories of both $\mathrm{EAR}_S^{\mathrm{cur}}$ and $\mathrm{EAR}_S^{\mathrm{pcur}}$.

(IV) We assume the definition of "regular monocursive S-function" given in §4.8 (Definitions 2 and 3) to be in the <u>metatheories</u> of EAR_S^{cur} and EAR_S^{pcur}. In turn, we assume the definitions of both "partially regular monocursive S-function" and "strictly partially regular monocursive S-function" also given in §4.8 (Definitions 2 and 3) to be in the <u>metatheory</u> of EAR_S^{pcur}.

(V) We assume the Gödel numbering of arithmetic RA_S^{mon} mentioned in Definition 2 of §4.8 to be in the <u>metatheories</u> of EAR_S^{cur} and EAR_S^{pcur}.

(VI) We assume the primitive recursive function $\rho_S(n, i)$ mentioned in Definition 2 of §4.8 that enumerates the Gödel numbers of $(n+1)$-place <u>regular</u> monocursive S-functions to be in the <u>metatheories</u> of EAR_S^{cur} and EAR_S^{pcur}. In turn, we assume the primitive recursive function $\rho'_S(n, i)$ mentioned in Definition 3 of §4.8 that enumerates the Gödel numbers of $(n+1)$-place <u>strictly partially regular</u> monocursive S-functions to be in the <u>metatheory</u> of EAR_S^{pcur}.

(VII) The FORMATION RULES for formulas of EAR_S^{mon} are as follows:

(1) An individual variable is a <u>term</u>.

(2) A constant in S is a <u>term</u>.

(3) For any $n \geq 1$, any n-place monocursive S-function

f, and any terms $\mathbf{t}_1$, $\mathbf{t}_2$, ..., $\mathbf{t}_n$, the expression $\mathbf{f}(\mathbf{t}_1,\mathbf{t}_2,\ldots,\mathbf{t}_n)$ is a term.

(4) For any terms $\mathbf{t}_1$ and $\mathbf{t}_2$, the expression $\mathbf{t}_1 = \mathbf{t}_2$ is a wff, with such a wff being called an equation.

(5) An expression is a term or wff (of $\mathrm{EAR}_S^{\mathrm{mon}}$) if and only if it is so by virtue of formation rules (1)-(4) above.

(VIII) The FORMATION RULES for formulas of $\mathrm{EAR}_S^{\mathrm{cur}}$ are as follows:

(1′) An individual variable is a term.

(2′) A constant in S is a term.

(3′) For any $n \geq 1$, any n-place monocursive S-function **f**, and any terms $\mathbf{t}_1$, $\mathbf{t}_2$, ..., $\mathbf{t}_n$, the expression $\mathbf{f}(\mathbf{t}_1,\mathbf{t}_2,\ldots,\mathbf{t}_n)$ is a term.

(4′) For any $n \geq 1$, any $i \geq 0$, and any terms $\mathbf{t}_1$, $\mathbf{t}_2$, ..., $\mathbf{t}_n$, the expression $f_i^n(\mathbf{t}_1,\mathbf{t}_2,\ldots,\mathbf{t}_n)$ is a term.

(5′) For any terms $\mathbf{t}_1$ and $\mathbf{t}_2$, the expression $\mathbf{t}_1 = \mathbf{t}_2$ is a wff, with such a wff being called an equation.

(6′) An expression is a term or wff (of $\mathrm{EAR}_S^{\mathrm{cur}}$) if and only if it is so by virtue of formation rules (1′)-(5′) above.

(IX) The FORMATION RULES for formulas of $\mathrm{EAR}_S^{\mathrm{pcur}}$ are as follows:

(1″) An individual variable is a <u>term</u>.

(2″) A constant in S is a <u>term</u>.

(3″) For any $n \geq 1$, any n-place monocursive S-function $\mathbf{f}$, and any terms $\mathbf{t}_1$, $\mathbf{t}_2$, ..., $\mathbf{t}_n$, the expression $\mathbf{f}(\mathbf{t}_1,\mathbf{t}_2,\ldots,\mathbf{t}_n)$ is a <u>term</u>.

(4″) For any $n \geq 1$, any $i \geq 0$, and any terms $\mathbf{t}_1$, $\mathbf{t}_2$, ..., $\mathbf{t}_n$, each of the expressions $f_i^n(\mathbf{t}_1,\mathbf{t}_2,\ldots,\mathbf{t}_n)$ and $g_i^n(\mathbf{t}_1,\mathbf{t}_2,\ldots,\mathbf{t}_n)$ is a <u>term</u>.

(5″) For any terms $\mathbf{t}_1$ and $\mathbf{t}_2$, the expression $\mathbf{t}_1 = \mathbf{t}_2$ is a <u>wff</u>, with such a wff being called an <u>equation</u>.

(6″) An expression is a term or wff (of $\mathrm{EAR}_S^{\mathrm{pcur}}$) if and only if it is so by virtue of formation rules (1″)-(5″) above.

(X) For $*** \in \{\mathrm{mon},\mathrm{cur},\mathrm{pcur}\}$, the AXIOMS and AXIOM SCHEMAS of EAR_S^{***} are the following:

(1) All quantifier-free propositions of semiological number theory concerning the semiological space S and placed in the equation-calculus form (see Goodstein [**1957**, Chapter III] and §4.10 above) are axioms of EAR_S^{***} (see (I) above).

(2) All defining equations of monocursive S-functions (including the projection and constant functions of S) are axioms of EAR_S^{***} (see (II) above).

(3) The following wff, which is a theorem of $\mathcal{F}_{\mathbf{BS}}$ (see Bourbaki [**1968**, p. 46, Theorem 1]), is an axiom of EAR_S^{***}:

AXIOM 1. $x = x$.

(4) We have the following two AXIOM SCHEMAS for $\mathrm{EAR}_S^{\mathrm{cur}}$ and $\mathrm{EAR}_S^{\mathrm{pcur}}$:

AXIOM SCHEMA I. For any $n \geq 1$ and $i \geq 0$, if γ_S is the regular monocursive S-function with Gödel number $\rho_S(n, i)$, then the following equation is an axiom of $\mathrm{EAR}_S^{\mathrm{cur}}$ and of $\mathrm{EAR}_S^{\mathrm{pcur}}$:

$$\gamma_S(x_1, \ldots, x_n, f_i^n(x_1, \ldots, x_n)) = \#_S.$$

AXIOM SCHEMA II. For any $n \geq 1$ and $i \geq 0$, if γ_S is the regular monocursive S-function with Gödel number $\rho_S(n, i)$, then the following equation is an axiom of $\mathrm{EAR}_S^{\mathrm{cur}}$ and of $\mathrm{EAR}_S^{\mathrm{pcur}}$ (see Proposition 14 of §4.10):

$$\mathbf{imp}_S(\overline{\mathrm{sg}}_S(|z, f_i^n(X_n)|_S), \overline{\mathrm{sg}}_S(|\gamma_S(X_n, z), \#_S|_S)) = \#_S.$$

(4) In the case of $\mathrm{EAR}_S^{\mathrm{pcur}}$ we have the following two AXIOM SCHEMAS:

AXIOM SCHEMA III. For any $n \geq 1$ and $i \geq 0$ and any constant terms $a_1, \ldots, a_n \in S$, if γ_S is the strictly partially regular monocursive S-function with Gödel number $\rho'_S(n, i)$ and if the relation

$$(\exists y)(\gamma_S(a_1, \ldots, a_n, y) = \#_S)$$

is a theorem of $\mathrm{RA}_S^{\mathrm{mon}}$, then the following equation is an axiom of $\mathrm{EAR}_S^{\mathrm{pcur}}$:

$$\gamma_S(a_1, \ldots, a_n, g_i^n(a_1, \ldots, a_n)) = \#_S.$$

AXIOM SCHEMA IV. For any $n \geq 1$ and $i \geq 0$ and any constant terms $a_1, \ldots, a_n \in S$, if γ_S is the strictly partially regular monocursive S-function with Gödel number $\rho'_S(n, i)$ and if the relation

$$(\exists y)(\gamma_S(a_1, \ldots, a_n, y) = \#_S)$$

is a theorem of $\mathrm{RA}_S^{\mathrm{mon}}$, then the following relation is an axiom of $\mathrm{EAR}_S^{\mathrm{pcur}}$:

$$\mathbf{imp}_S(\overline{\mathrm{sg}}_S(|z, g_i^n(X_n)|_S), \overline{\mathrm{sg}}_S(|\gamma_S(X_n, z), \#_S|_S)) = \#_S.$$

(XI) The following are INFERENCE RULES of $\mathrm{EAR}_S^{\mathrm{mon}}$, $\mathrm{EAR}_S^{\mathrm{cur}}$, and $\mathrm{EAR}_S^{\mathrm{pcur}}$:

(1) The following two inference rules, which are derivable in $\mathcal{F}_{\mathrm{BS}}$ (see Bourbaki [1968, p. 46,

Theorems 2 and 3]):

$$(E_1) \qquad \frac{T = U}{U = T}$$

$$(E_2) \qquad \frac{\begin{array}{c} T = U \\ U = V \end{array}}{T = V}$$

(2) The following inference rule is derivable in a straightforward manner from Criterion C61 (The Principle of Induction) of Bourbaki [1968, p. 168]:

$$(\mathrm{IND}) \qquad \frac{\begin{array}{ll} f(0) = g(0) & \\ f(\sigma_i(x)) = H_{\sigma_i(\#_S)}(x, f(x)) & (\sigma_i \in A_S) \\ g(\sigma_i(x)) = H_{\sigma_i(\#_S)}(x, g(x)) & (\sigma_i \in A_S) \end{array}}{f(x) = g(x)}$$

(3) EAR_S^{***} ($*** \in \{\mathrm{mon}, \mathrm{cur}, \mathrm{pcur}\}$) has the following two inference rules, the first of which is a special case of Criterion C3 of Bourbaki [1968, p. 26] and the second of which is Criterion C44 of Bourbaki [1968, p. 46]:

$$(S_1) \qquad \frac{\mathbf{f}(x) = \mathbf{g}(x)}{\mathbf{f}(T) = \mathbf{g}(T)}$$

$$(S_2) \qquad \frac{T = U}{\mathbf{f}(T) = \mathbf{f}(U)}$$

(4) In addition, in EAR_S^{cur} and EAR_S^{pcur} we have the following <u>Uniqueness Rule</u>:

$$\mathbf{E}_1, \ldots, \mathbf{E}_k$$

$$\mathbf{E}_k \text{ is the formula } (\#_S|x)\mathbf{E}_{k+1}$$

$$(U_1) \qquad \frac{\mathbf{E}_{k+1} \vdash_{1,2} \mathbf{E}_{k+2}^{(i)}, \ldots, \mathbf{E}_{k+n_i-1}^{(i)}, (\sigma_i(x)|x)\mathbf{E}_{k+1} \quad (\sigma_i \in A_S)}{\mathbf{E}_{k+1}}$$

where the $\mathbf{E}_i$ are equations, x is an individual variable, and $\vdash_{1,2}$ indicates a derivation by means of inference rules (S_1) and (S_2). The equations

$$\mathbf{E}_{k+1}^{(i)}, \ldots, \mathbf{E}_{k+n_i-1}^{(i)}, (\sigma_i(x)|x)\mathbf{E}_{k+1} \qquad (\sigma_i \in A_S)$$

are called <u>conditional equations</u> (cf. footnote ([2]) to Uniqueness Rule (U) in item (4) of part (XI) of §4.4).

(X) For *** ∈ {mon,cur,pcur}, no wff of $\mathscr{F}_{\mathbf{BS}}$ is a theorem of EAR_S^{***} unless it is derivable from the axioms and axiom schemas of EAR_S^{***} using the inference rules of EAR_S^{***}. We let $Wff(EAR_S^{***})$ denote the set of <u>wffs</u> of EAR_S^{***} and let $Thm(EAR_S^{***})$ denote the set of <u>theorems</u> of EAR_S^{***}.

§4.12. Throughout this section we let S be a zeroed semiological space with successor set

$$A_S = \{\sigma_1, \sigma_2, \sigma_3, \ldots\},$$

where it is permitted that A_S be either finite (but nonempty) or denumerably infinite, and with root set $R_S = \{\#_S\}$. We let ζ denote the zeroed successor function of space S. Without loss of generality, we may let σ_1 be a successor function such that $\sigma_1(\#_S) \notin R_S$ (Pogorzelski and Ryan [1982, p. I.1.1.1, Axiom PS5]), in which case $\sigma_1(\#_S) \neq \#_S$. For convenience, we denote the term $\sigma_1(\#_S)$ as 1_S, so that we have

$$(*) \qquad 1_S = \sigma_1(\#_S), \qquad 1_S \in S, \qquad 1_S \neq \#_S.$$

Then, we see (cf. (*) in the paragraph following Proposition 1 of §4.10) that both equations of the monocursive scheme define the function f at $\#_S$; specifically, we have

$$f(X_n,\#_S) = g(X_n)$$

$$f(X_n,\#_S) = f(X_n,\zeta(\#_S)) = h_{\zeta(\#_S)}(X_n,\#_S,g(X_n)).$$

Thus (see Proposition 4 of §4.10), given such a zeroed semiological space S, the monocursive scheme defines a function if and only if we have

$$h_{\zeta(\#_S)}(X_n,\#_S,g(X_n)) = g(X_n).$$

Thus, in the case where a semiological space S is <u>zeroed</u>, in attempting to define a function f by means of the monocursive scheme, one sees that in a great many cases the function $h_{\zeta(\#_S)}$ that gives the desired values for <u>nonroots</u> does not give the same value for $\zeta(\#_S)$ (= $\#_S$) as does the <u>first</u> equation of the monocursive scheme. In fact, we make the following conjecture:

CONJECTURE 1. For any zeroed semiological space S, with root $\#_S$, there does not exist a monocursive function $|x,y|_S$ such that each of the relations

$$|x,y|_S = \#_S \qquad \text{and} \qquad x = y$$

is derivable from the other in the sentential [equational] arithmetic SAR_S^{mon} [EAR_S^{mon}].

However, if the monocursive scheme (Pogorzelski and Ryan [1985, pp. 0.4.3.1-2]) were replaced by the following <u>modified monocursive scheme</u>:

$$(**) \quad \begin{aligned} f(X_n, \#_S) &= g_\#(X_n) \qquad (\# \in R_S) \\ f(X_n, \sigma(y)) &= h_{\sigma(\#)}(X_n, y, f(X_n, y)) \\ &\qquad (\sigma(\#) \in A_S^*,\ \sigma(y) \notin R_S) \ ([1]), \end{aligned}$$

where $(g_\#)_{\# \in R_S}$ and $(h_{\sigma(\#)})_{\sigma(\#) \in A_S^*}$ are families satisfying the conditions given in Pogorzelski and Ryan [1985, p. 0.4.3.1], then the following results are obtained:

([1]) That is, the condition $\sigma(y) \notin R_S$ is added to the second equation of the monocursive scheme, eliminating the "overlapping" domain of definition of the two equations of the monocursive scheme.

PROPOSITION 1. _If_ $(g_{\#})_{\# \in R_S}$ _and_ $(h_{\sigma(\#)})_{\sigma(\#) \in A_S^*}$ _are families of total_ n_-place_ cl(S)_-functions and_ $(n+2)$_-place_ cl(S)_-functions, respectively, then the modified monocursive scheme_ (**) _defines a total_ $(n + 1)$_-place_ cl(S)_-function._

Now, recalling that we are assuming a _zeroed_ semiological space S and are assuming the _modified_ monocursive scheme (**), we define the following functions and obtain the following results [2,3]:

DEFINITION 1. We define a function $x \oplus y$, which we call _addition_, by the modified monocursive scheme:

$$x \oplus \#_S = x$$

$$x \oplus \sigma(y) = \sigma(x \oplus y) \quad (\sigma(y) \notin R_S).$$

PROPOSITION 2. $\#_S \oplus x = x$.

---

(2) In the following we use the modified monocursive scheme as it would apply in the case of _semiological_ spaces: namely, where space S is closed (so that S = cl(S)) and singly rooted, so that instead of a family $(g_{\#})_{\# \in R_S}$ we simply have a single function g. Moreover, all of the successor functions of space S are total S-functions. In addition, in most of the following definitions the family $(h_{\sigma(\#_S)})_{\sigma(\#_S) \in A_S^*}$ is such that all of the functions $h_{\sigma(\#_S)}$ are identical; that is, we do not deal with a _family_, but only a single function h.

(3) Our proofs and sequence of definitions and propositions follow Vučković [1962, §§2-3].

Proof. By presemiological induction using Definition 1.

We point out that our proofs in this section are in the form of proofs in a *sentential* calculus. However, we make the following conjecture:

CONJECTURE 2. We make the conjecture that the proofs in this section can be transcribed into proofs in an *equational* calculus in a straightforward manner.

We now have the following derived inference rule:

DERIVED INFERENCE RULE I.

$$\frac{X \oplus Y = \#_S}{X = \#_S, \ Y = \#_S}$$

Proof. By presemiological induction using Definition 1.

LEMMA (for Proposition 3). $\sigma(z) \notin R_S \rightarrow \sigma(y \oplus z) \notin R_S$.

Proof. One proves the contrapositive of the Lemma by means of Derived Inference Rule I.

PROPOSITION 3. $(x \oplus y) \oplus z = x \oplus (y \oplus z)$.

Proof. By presemiological induction using Definition 1. When applying the induction hypothesis, one uses the Lemma.

DEFINTION 2. We define a function $x \odot y$, which we call *multiplication*, by the modified monocursive scheme:

$$x \odot \#_S = \#_S$$

$$x \odot \sigma(y) = (x \odot y) \oplus x \qquad (\sigma(y) \notin R_S).$$

PROPOSITION 4. <u>For</u> $\sigma(x) \notin R_S$, <u>we have</u>

$$\sigma(x) \neq \#_S \Rightarrow x \odot \sigma(\#_S) = x.$$

<u>In particular, we have</u> $x \odot \sigma_1(\#_S) = x \odot \mathbf{1}_S = x$.

<u>Proof</u>. One uses Definition 2, Proposition 2, and the relations $\mathbf{1}_S = \sigma_1(\#_S)$ and $\mathbf{1}_S \neq \#_S$ in (*) in the first paragraph of this subsection.

PROPOSITION 5. $x \odot (y \oplus z) = (x \odot y) \oplus (x \odot z)$.

<u>Proof</u>. By presemiological induction using Definitions 1 and 2. In applying the induction hypothesis one uses the Lemma for Proposition 3, as well as Proposition 3, itself.

PROPOSITION 6. $\#_S \odot x = \#_S$.

<u>Proof</u>. By presemiological induction, using Definitions 2 and 1.

PROPOSITION 7. $x \odot (y \odot z) = (x \odot y) \odot z$.

<u>Proof</u>. By presemiological induction, using Definition 2. In applying the induction hypothesis, one also uses Proposition 5.

DEFINITION 3. We now define a function, which we denote as $x \ominus y$ and call <u>restricted subtraction</u>, with $x \ominus y$ being defined by double recursion (cf. Vučković [1962, §3, definition of $X \dot{-} Y$]):

$$x \ominus \#_S = x$$

$$\#_S \ominus \sigma(y) = \#_S \qquad (\sigma(y) \notin R_S)$$

$$\sigma(x) \ominus \tau(y) = \begin{cases} x \ominus y & (\sigma = \tau,\ \sigma(x) \notin R_S,\ \tau(y) \notin R_S) \\ \sigma(x \ominus y) & (\sigma \neq \tau,\ \sigma(x) \notin R_S,\ \tau(y) \notin R_S). \end{cases}$$

One can easily provide an inductive proof that this definition does, in fact, define a total 2-place S-function. Inasmuch as we will not be concerned with arithmetics over zeroed semiological spaces in this volume, we leave it as an open question whether or not our function of restricted subtraction can be defined by means of the modified monocursive scheme, given above. However, we make the following conjecture (cf. the assertion of Vučković [1962, p. 147] that his function $X \dot{-} Y$ is primitive recursive):

CONJECTURE 3. We conjecture that the function of restricted subtraction $x \ominus y$ can, in fact, be defined using the modified monocursive scheme, given above.

PROPOSITION 8. $\#_S \ominus x = \#_S$.

<u>Proof</u>. Consider cases $x = \#_S$ and $x \neq \#_S$, using Definition 3.

PROPOSITION 9. $x \ominus (y \oplus x) = \#_S$.

Proof. By presemiological induction using Propsition 8 for the case $x = \#_S$, and using Definitions 1 and 3 and the Lemma for Proposition 3 when applying the induction hypothesis.

PROPOSITION 10. $(y \oplus x) \ominus x = y$.

Proof. By presemiological induction using Definitions 3 and 1 and the Lemma for Proposition 3.

PROPOSITION 11. $(z \oplus x) \ominus (y \oplus x) = z \ominus y$.

Proof. By presemiological induction using Definitions 1 and 3 and the Lemma for Proposition 3.

PROPOSITION 12. $x \ominus x = \#_S$.

Proof. By presemiological induction using Definition 3.

Recalling our notational convention of defining $1_S = \sigma_1(\#_S)$, we make the following definition:

DEFINITION 4. We define a function, which we denote as $\overline{sg}_S(x)$ and call the signum bar function, by the modified monocursive scheme:

$$\overline{sg}_S(\#_S) = 1_S$$

$$\overline{sg}_S(\sigma(x)) = \#_S \quad (\sigma(x) \notin R_S).$$

PROPOSITION 13. <u>For all</u> $x \in S$, <u>we have</u>

$$\overline{\mathrm{sg}}_S(x) = \#_S \iff x \neq \#_S.$$

<u>Proof</u>. By Definition 4.

DEFINITION 5. We define a function, which we denote as $\mathrm{sg}_S(x)$ and call the <u>signum function</u>, by the modified monocursive scheme:

$$\mathrm{sg}_S(\#_S) = \#_S$$

$$\mathrm{sg}_S(\sigma(x)) = 1_S \quad (\sigma(x) \notin R_S).$$

PROPOSITION 14. <u>For all</u> $x \in S$, <u>we have</u>

$$\mathrm{sg}_S(x) = \#_S \iff x = \#_S.$$

<u>Proof</u>. By Definition 5.

PROPOSITION 15. $\overline{\mathrm{sg}}_S(x) = 1_S \ominus \mathrm{sg}_S(x)$.

<u>Proof</u>. In the case where $x = \#_S$ one uses Definitions 4, 3, and 5. In the case where $x \neq \#_S$ one uses Definition 4, Proposition 12, and Definition 5.

PROPOSITION 16. $x \odot \overline{\mathrm{sg}}_S(x) = \#_S$.

<u>Proof</u>. In the case where $x = \#_S$ one uses Proposition 6, and in the case where $x \neq \#_S$ one uses Definitions 4 and 2.

DEFINITION 6. We define a function, which we denote as $|x,y|_S$ and call the <u>absolute difference</u>, by:

$$|x,y|_S = (x \ominus y) \oplus (y \ominus x).$$

PROPOSITION 17. <u>For any</u> x, $y \in S$ <u>and</u> $\sigma \in A_S$ <u>such that</u> $\sigma \neq \zeta$,

$$\sigma(y) \odot \overline{\mathrm{sg}}_S(x) = (y \odot \overline{\mathrm{sg}}_S(x)) \oplus (\sigma(\#_S) \odot \overline{\mathrm{sg}}_S(x)).$$

<u>Proof</u>. In the case where $x = \#_S$ one uses Definition 4, Proposition 4, and Definition 1. In the case where $x \neq \#_S$ one uses Proposition 13 and Definitions 2 and 1.

Now we point out that the proof of Vučković [1962, p. 148] of his equation (3.20), which in our notation would be written in the form

$$(\dagger) \qquad y \odot \overline{\mathrm{sg}}_S(x \ominus y) = ((y \ominus x) \oplus x) \odot \overline{\mathrm{sg}}_S(x \ominus y),$$

does not carry over to the case where the underlying semiological space is zeroed; i.e., to the case of an arithmetic using the modified monocursive scheme and over a zeroed semiological space, such as we have been considering in the present subsection.

In his above-mentioned proof of his equation (3.20), Vučković, setting

$$f(X,Y) = Y \cdot \{e \dot{-} \alpha(X \dot{-} Y)\},$$

obtains the equation

(††) $f(a_\xi X, a_\eta Y) = (a_\xi Y)\cdot\{e \dot{-} \alpha(X \dot{-} Y)\}$

$= f(X,Y) + a_\xi\cdot\{e \dot{-} \alpha(X \dot{-} Y)\},$

for $a_\xi = a_\eta$,

which in our notation would be written

(†††) $\sigma(y) \odot \overline{sg}_S(\tau(x) \ominus \sigma(y)) =$

$(y \odot \overline{sg}_S(x \ominus y)) \oplus (\sigma(\#_S) \odot \overline{sg}_S(x \ominus y))$

for $\sigma = \tau$ (4).

However, in the case where

$$\sigma = \tau = \zeta \qquad \text{and} \qquad x = y = 1_S \;(= \sigma_1(\#_S))$$

the equation (†††) would yield

$$\zeta(1_S) \odot \overline{sg}_S(\zeta(1_S) \ominus \zeta(1_S)) =$$
$$(1_S \odot \overline{sg}_S(1_S \ominus 1_S)) \oplus (\zeta(\#_S) \odot \overline{sg}_S(1_S \ominus 1_S)),$$

which with

$\overline{sg}_S(\zeta(1_S) \ominus \zeta(1_S)) = \overline{sg}_S(\#_S)$ [Proposition 12]

$= 1_S$ [Definition 4],

and, similarly,

---

(4) We note (cf. Vučković [1959, §2, third paragraph]) that in the Vučković equation (††) above, the term a_ξ (which is a letter of Vučković's alphabet $\mathfrak{A}$) occurring immediately before the multiplication sign "·" would correspond to a term of the form $a_\xi(0)$ in an arithmetic using successor functions instead of letters of an alphabet.

$$\overline{sg}_S(1_S \ominus 1_S) = \overline{sg}_S(\#_S) = 1_S,$$

along with $\zeta(\#_S) = \#_S$, would yield

$$\zeta(1_S) \odot 1_S = (1_S \odot 1_S) \oplus (\#_S \odot 1_S),$$

which with Proposition 4 and Definition 1 would yield

$$\zeta(1_S) = 1_S \oplus \#_S = 1_S,$$

which with $1_S = \sigma_1(\#_S)$ would yield

$$(\dagger\dagger\dagger\dagger) \qquad \zeta(\sigma_1(\#_S)) = \sigma_1(\#_S).$$

But with $\sigma_1(\#_S) \neq \#_S$ (see the first paragraph of this section) it follows (see Pogorzelski and Ryan [**1982**, Chapter I, §1, no. 4, Definition 1, Axiom ZPS1]) that (††††) is, in fact, false.

Thus by contradiction we have shown that (if our arithmetics are consistent) Vučković's proof of his equation (3.20) mentioned above does not carry over to the case where the underlying semiological space is zeroed (and where our modified monocursive scheme is used).

However, we make the following conjecture:

CONJECTURE 4. The equation (†) above, which corresponds to Vučković's equation (3.20) mentioned above, is provable in any of our sentential or equational arithmetics over zeroed semiological spaces, where the arithmetics use our modified monocursive scheme.

If Conjecture 4 is true, i.e., if equation (†) above is a theorem of the arithmetic we are constructing over some zeroed semiological space using the modified monocursive scheme, then from (†), along with Definition 4 and Proposition 4, we would obtain the following derived inference rule:

DERIVED INFERENCE RULE II.

$$\frac{X \ominus Y = \#_S}{Y = (Y \ominus X) \oplus X}$$

In turn (still assuming Conjecture 4 to be true), one could obtain the following derived inference rule:

DERIVED INFERENCE RULE III. Each of the relations $|X,Y|_S = \#_S$ and $X = Y$ can be derived from the other.

Proof. When the hypothesis is $X = Y$, one uses Definition 6, Proposition 12, and Definition 1. When the hypothesis is $|X,Y|_S = \#_S$, one uses Definition 6, Derived Inference Rules I and II, and Proposition 2.

Thus, with the material in the present section, along with Propositions 5-9 of §4.10, we see that we have proved that, if Conjectures 2-4 are true, then the propositional calculus with equality can be developed in an equational arithmetic over a zeroed semiological space, provided the arithmetic uses the modified monocursive scheme. Moreover, if Conjecture 1 is true, we cannot develop the monocursive function that corresponds to equality in an equational arithmetic over a zeroed semiological space, unless the arithmetic uses the modified monocursive scheme.

§4.13. For presemiological spaces S and S', let $\alpha SS'$ be a semicarrier (in which case $\alpha: S \longrightarrow S'$ is a partial function). We see that if for some x, $y \in S$ we had $(x,y) \in S$, then there would exist a 1-place S-function f that is also a 2-place S-function, so that in certain instances the value $\alpha[f]$ would not be uniquely determined (see Pogorzelski and Ryan [**1982**, Chapter 0, §1, no. 3, especially Remark 1] for details and an example). It was for this reason that we introduced the formal notation $\alpha^{(n)}[f]$ for the extension of a mapping α to functions, where, in the extension of α that is being considered, the functions f are to be considered as n-place S-functions (Pogorzelski and Ryan [**1982**, pp. 0.1.3.1-0.1.3.2]).

However, we see that if space S is <u>atomic</u> (Pogorzelski and Ryan [**1985**, pp. 0.4.6.19-20]), then the situation of a nonempty function f being both an n-place S-function and an m-place S-function for $n \neq m$ cannot arise, so that, provided the value $\alpha^{(n)}[\emptyset]$ is defined to be the same for all n (in the event that the empty function $\emptyset$ belongs to the set of S-functions f for which $\alpha^{(n)}[f]$ is being considered), the inclusion of the number n of places of the function f in the notation $\alpha^{(n)}[f]$ is no longer necessary (see Pogorzelski and Ryan [**1985**, p. 0.4.1.21, part (c) of Proposition 13]).

Therefore, when considering semicarriers $\alpha SS'$, in cases where we want to consider extending α to S-functions, we will require that the presemiological space S be <u>atomic</u>, thus eliminating the need for considering an infinite sequence of semicarriers $(\alpha^{(n)}SS')_{n \in \mathbf{N}^*}$.

Similarly, when considering carriers $\alpha SS'\beta$, in cases where we want to consider extending α and β to S-functions and S'-functions, respectively, we will require that the presemiological spaces S and S' be atomic.

We note that G. Asser [1960] in dealing with recursive word functions does not have the above-mentioned problem of a function being both n-place and m-place for different n and m. This is due to the fact that a concatenation of n letters $a_1a_2\ldots a_n$, for $n \geq 1$, is considered to be different from the n-tuple of letters $(a_1,a_2,\ldots,a_n)$.

For example, if f is a 1-place word function and is defined at the word $a_1a_2a_3$, then f will not be defined at the ordered pair of words $((a_1a_2),a_3)$ nor at the ordered triple of words (a_1,a_2,a_3).

On the other hand, inasmuch as we are working within Bourbaki's theory of sets, our analog of a concatenation of letters $a_1a_2\ldots a_n$ $(n \geq 1)$ is, in fact, an n-tuple $(a_1,a_2,\ldots,a_n)$, which is also an ordered pair $((a_1,a_2,\ldots,a_{n-1}),a_n)$ (if $n \geq 2$), an ordered triple

$$((a_1,a_2,\ldots,a_{n-2}),a_{n-1},a_n)$$

(if $n \geq 3$), ..., and an ordered $(n-1)$-tuple $((a_1,a_2),a_3,\ldots,a_n)$ (if $n \geq 3$).

Therefore this problem, which was nonexistent for Asser, is a problem for us.

In the remainder of this section we present Asser's results (in our notation and terminology):

Let S be an atomic finite-dimensional nonzeroed noncommutative semiological space, with root set $R_S = \{\#_S\}$ and successor set $A_S = \{\sigma_1, \sigma_2, \ldots, \sigma_n\}$. Then Asser [1960] defines a carrier $\mathbf{g}_S S E \mathbf{g}_S^{-1}$ by defining a function $\mathbf{g}_S : S \longrightarrow E$ by:

$$\mathbf{g}_S(\#_S) = 0 \; (= \#_E)$$

$$\mathbf{g}_S(\sigma_{i_\ell} \circ \ldots \circ \sigma_{i_1}(\#_S)) = \sum_{p=1}^{\ell} i_p n^{p-1}.$$

One sees that $\mathbf{g}_S : S \longrightarrow E$ is a bijection, so that the inverse mapping $\mathbf{g}_S^{-1} : E \longrightarrow S$ exists.

Asser [1960, §§3, 5, Main Theorems] then proves the following results:

THEOREM (Asser). Let S be an atomic finite-dimensional nonzeroed noncommutative semiological space, and let $*** \in \{\text{mon}, \text{cur}, \text{pcur}\}$. Then we have $\mathbf{g}_S\langle \mathbf{F}_S^{***} \rangle = \mathbf{F}_E^{***}$ [1,2].

One can extend $\mathbf{g}_S$ to a mapping

$$\bar{\mathbf{g}}_S : \mathbf{Wff}(\text{SAR}_S^{***}) \longrightarrow \mathbf{Wff}(\text{SAR}_E^{***}) \; [3]$$

---

(1) We note that in the case of the Peano semiological space E we have $*** \in \{\text{pre}, \text{rec}, \text{parec}\}$.

(2) The notation $G\langle X \rangle$ $(= \{G(x) \mid x \in X\})$ is introduced in Bourbaki [1968, p. 77, Definition 3].

(3) We note that for any function $\mathbf{g}$, Asser uses the notation $\bar{\mathbf{g}}$ to denote the inverse function $\mathbf{g}^{-1}$.

by defining for each $\mathbf{A} \in \mathbf{Wff}(\mathrm{SAR}_S^{***})$, the value $\bar{\mathbf{g}}_S(\mathbf{A})$ to be the formula obtained by

(i) replacing each occurrence in $\mathbf{A}$ of a resolution

$$(*) \qquad \sigma_{i_\ell}\circ\ldots\circ\sigma_{i_1}(\#_S) \qquad (\ell \geq 1)$$

by the value $\mathbf{g}_S(\sigma_{i_\ell}\circ\ldots\circ\sigma_{i_1}(\#_S))$;

(ii) replacing each occurrence in $\mathbf{A}$ of the root $\#_S$ that is not an occurrence in a resolution of the form (*) by 0;

(iii) replacing each occurrence in $\mathbf{A}$ of a function f that is not an occurrence in a resolution of the form (*) by the function $\mathbf{g}_S[f]$.

From $\mathbf{g}_S : S \longrightarrow E$ being a bijection it follows that

$$\bar{\mathbf{g}}_S : \mathbf{Wff}(\mathrm{SAR}_S^{***}) \longrightarrow \mathbf{Wff}(\mathrm{SAR}_E^{***})$$

is also a bijection, so that the inverse mapping

$$\bar{\mathbf{g}}_S^{-1} : \mathbf{Wff}(\mathrm{SAR}_E^{***}) \longrightarrow \mathbf{Wff}(\mathrm{SAR}_S^{***})$$

exists.

In turn, one can extend $\mathbf{g}_S$ to a mapping

$$\bar{\bar{\mathbf{g}}}_S : \mathbf{Wff}(\mathrm{EAR}_S^{***}) \longrightarrow \mathbf{Wff}(\mathrm{EAR}_E^{***})$$

by defining for each $\mathbf{A} \in \mathbf{Wff}(\mathrm{EAR}_S^{***})$, the value $\bar{\bar{\mathbf{g}}}_S(\mathbf{A})$ to be the formula obtained by

(i') replacing each occurrence in $\mathbf{A}$ of a resolution

(**) $\sigma_{i_\ell}\circ\ldots\circ\sigma_{i_1}(\#_S)$ $(\ell \geq 1)$

by the value $\mathbf{g}_S(\sigma_{i_\ell}\circ\ldots\circ\sigma_{i_1}(\#_S))$;

(ii') replacing each occurrence in **A** of the root $\#_S$ that is not an occurrence in a resolution of the form (**) by 0;

(iii') replacing each occurrence in **A** of a function f that is not an occurrence in a resolution of the form (**) by the function $\mathbf{g}_S[f]$.

From $\mathbf{g}_S: S \longrightarrow E$ being a bijection it follows that

$$\bar{\bar{\mathbf{g}}}_S: \mathrm{Wff}(\mathrm{EAR}_S^{***}) \longrightarrow \mathrm{Wff}(\mathrm{EAR}_E^{***})$$

is also a bijection, so that the inverse mapping

$$\bar{\bar{\mathbf{g}}}_S^{-1}: \mathrm{Wff}(\mathrm{EAR}_E^{***}) \longrightarrow \mathrm{Wff}(\mathrm{EAR}_S^{***})$$

exists.

We then have the following proposition:

PROPOSITION 1. _For any atomic finite-dimensional nonzeroed noncommutative semiological space S and for_

$$*** \in \{\mathrm{mon}, \mathrm{cur}, \mathrm{pcur}\},$$

we have

$$\bar{\mathbf{g}}_S\langle \mathrm{Wff}(\mathrm{SAR}_S^{***})\rangle = \mathrm{Wff}(\mathrm{SAR}_E^{***}),$$

$$\bar{\bar{\mathbf{g}}}_S\langle \mathrm{Wff}(\mathrm{EAR}_S^{***})\rangle = \mathrm{Wff}(\mathrm{EAR}_E^{***}).$$

COROLLARY. For any atomic finite-dimensional nonzeroed noncommutative semiological space S and for $*** \in \{mon, cur, pcur\}$, we have

$$\bar{\mathbf{g}}_S\langle Thm(SAR_S^{***})\rangle = Thm(SAR_E^{***}),$$

$$\bar{\bar{\mathbf{g}}}_S\langle Thm(EAR_S^{***})\rangle = Thm(EAR_E^{***}).$$

§4.14. In this section we carry over to infinite-dimensional semiological spaces and show some of the pathologies involved with respect to such spaces.

Recalling that for any semiological space S, the terms $\mathbf{F}_S^{mon}$, $\mathbf{F}_S^{cur}$, and $\mathbf{F}_S^{pcur}$ denote the monocursive, cursive, and partial cursive functional spaces, respectively, over S (Pogorzelski and Ryan [**1985**, pp. II.2.1.8, II.3.2.4, II.3.1.5]), we have the following proposition:

PROPOSITION 1. <u>For any infinite-dimensional semiological space</u> S (<u>noncommutative</u>, <u>commutative</u>, <u>or</u> <u>zeroed</u>), <u>each of</u>

$$\mathrm{Card}(\mathbf{F}_S^{mon}), \qquad \mathrm{Card}(\mathbf{F}_S^{cur}), \qquad \mathrm{Card}(\mathbf{F}_S^{pcur})$$

<u>is nondenumerably infinite</u>.

<u>Proof</u>. We first establish that $\mathrm{Card}(\mathbf{F}_S^{mon})$ is denumerably infinite in each of the three possible cases:

Case 1. In the case where space S is <u>noncommutative</u> (and nonzeroed) we see that $\mathrm{Card}(\mathbf{F}_S^{mon})$ is denumerably infinite by the result of Schwenkel [**1965**, §2, first paragraph].

Case 2. Assume that space S is <u>zeroed</u> (in which case it is not commutative). Let $A_S = \{\zeta, \sigma_1, \sigma_2, \sigma_3, \ldots\}$, where ζ is the unique zeroed successor function of space S (Pogorzelski and Ryan [**1982**, Chapter I, §4, no. 1, Definition 1]), let $R_S = \{\#_S\}$, and let $\mathbf{g}(x)$ be the 1-place constant S-function satisfying

$$\mathbf{g}(x) = \#_S \quad (x \in S).$$

For convenience in defining families, we set $\sigma_0 = \zeta$, so that we can also write $A_S = \{\sigma_0, \sigma_1, \sigma_2, \sigma_3, \ldots\}$.

Let $c_0, c_1, c_2, \ldots, c_{10}$ be the 3-place S-functions defined by:

$$c_0 = \zeta \circ pr_3^3,\ c_1 = \sigma_1 \circ pr_3^3,\ c_2 = \sigma_2 \circ pr_3^3,\ \ldots,\ c_{10} = \sigma_{10} \circ pr_3^3,$$

where pr_3^3 is the 3rd 3-place projection function on S (Pogorzelski and Ryan [**1985**, Chapter 0, §4, no. 1]). We see that $c_0, c_1, c_2, \ldots, c_{10}$ are distinct monocursive S-functions defined by the composition scheme from initial functions of space S (Pogorzelski and Ryan [**1985**, Chapter II, §2, no. 1, Definition 1 and Propositions 1 and 2]).

Then, let $\mathscr{K}$ denote the set of all families of the form

$$(*) \qquad (h_{\sigma_i(\#_S)})_{\sigma_i(\#_S) \in A_S^*}$$

of 3-place S-functions $h_{\sigma_i(\#_S)}$ satisfying the following three conditions:

(i) <u>We have</u> $h_{\sigma_0(\#_S)} = c_0 = \zeta \circ pr_3^3 = \sigma_0 \circ pr_3^3$.

(ii) <u>For each</u> $\sigma_i(\#_S) \in A_S^*$ $(i \geq 1)$ <u>we have</u>

$$h_{\sigma_i(\#_S)} \in \{c_1, c_2, \ldots, c_{10}\}.$$

(iii) <u>There does not exist a</u> $k \geq 1$ <u>such that for all</u> $i \geq k$ <u>we have</u> $h_{\sigma_i(\#_S)} = c_{10}$.

One sees that each family of the form (*) in $\mathscr{K}$, together with the function **g**, defines a 2-place monocursive function over space S by the monocursive scheme (with different families defining different monocursive functions) and that each function **f** defined by the function **g** and a family of the form (*) can be associated with the sequence of functions

$$\mathbf{h}_{\sigma_1(\#_S)}, \mathbf{h}_{\sigma_2(\#_S)}, \mathbf{h}_{\sigma_3(\#_S)}, \ldots$$

(noting that **g** and $\mathbf{h}_{\sigma_0(\#_S)}$ $(= \mathbf{h}_{\zeta(\#_S)} = \mathbf{h}_{\#_S})$ are the same for each of the functions **f** thus defined), which by condition (ii) can be associated with some sequence of functions

$$\mathbf{g}, \mathbf{c}_{k_1}, \mathbf{c}_{k_2}, \mathbf{c}_{k_3}, \ldots,$$

which, in turn, can be associated with the expression

$$0.d_1d_2d_3\ldots,$$

where for $i = 1, 2, 3, \ldots$, we have $d_i = k_i - 1$, so that d_i is one of the digits 0, 1, ..., 9, so that $0.d_1d_2d_3\ldots$ is the decimal expansion of a real number in the interval [0, 1]. Moreover, by (iii) above, the real number whose decimal expansion is $0.d_1d_2d_3\ldots$ is uniquely determined (cf. Mendelson [**1973**, p. 274, Theorem 9.11]).

Conversely, one easily sees that each decimal expansion

$$0.d_1d_2d_3\ldots$$

of a real number in the interval [0, 1] which does not, from some point on, consist of only the digit "9" repeated infinitely many times determines a unique family of the form (*) in $\mathfrak{X}$. (Specifically, one defines $h_{\sigma_0}(\#_S) = \zeta$ and for $i = 1, 2, 3, \ldots$ defines $h_{\sigma_i}(\#_S) = c_{d_i+1}$.)

One thus sees that there is a one-to-one correspondence between a certain subset of the set F_S^{mon} of monocursive functions over space S and the set of real numbers in the interval [0, 1]. This establishes that $\mathrm{Card}(F_S^{mon})$ is denumerably infinite in the case where S is zeroed.

Case 3. The case where space S is <u>commutative</u> is handled along the lines of Case 2, but more simply. In particular, there is no zeroed successor function ζ, one has

$$A_S = \{\sigma_1, \sigma_2, \sigma_3, \ldots\},$$

the function $c_0 = \zeta \circ pr_3^3$ is eliminated and the functions c_1, c_2, ..., c_{10} are relabeled c_0, c_1, ..., c_9, condition (i) is eliminated, in condition (ii) the relation $(i \geq 1)$ is eliminated, and in condition (iii) the notation c_{10} is changed to c_9.

Thus in any case $\mathrm{Card}(F_S^{mon})$ is denumerably infinite. The proof of the proposition is completed using the relation

$$F_S^{mon} \subseteq F_S^{cur} \subseteq F_S^{pcur}$$

(Pogorzelski and Ryan [**1985**, Chapter II, §3, no. 2, Proposition 5]).

REMARK 1. The difference between Proposition 1 and the result of Schwenkel [1965, §2, first paragraph] is that:

First, while Schwenkel proves that in the case of infinite-dimensional noncommutative semiological spaces all S-functions are monocursive, Proposition 1 asserts that in the case of all infinite-dimensional semiological spaces there are nondenumerably many monocursive functions over S.

Second, while Schwenkel's result holds only for infinite-dimensional noncommutative semiological spaces, Proposition 1 holds for all infinite-dimensional semiological spaces.

Third, while Schwenkel's result deals only with monocursive functions, Proposition 1 deals with monocursive, cursive, and partial cursive functions.

Then, since for every monocursive function $\mathbf{f}(x)$ of a semiological space S, the relation $\mathbf{f}(x) = \mathbf{f}(x)$ is a theorem of each sentential or equational monocursive, cursive, or partial cursive arithmetic of space S, we have the following proposition and corollary:

PROPOSITION 2. For any infinite-dimensional semiological space S and for *** $\in$ {mon, cur, pcur},

(a) $\mathrm{Card}(\mathbf{Thm}(\mathrm{SAR}_S^{***}))$ is nondenumerably infinite;

(b) $\mathrm{Card}(\mathbf{Thm}(\mathrm{EAR}_S^{***}))$ is nondenumerably infinite.

COROLLARY. _For any infinite-dimensional semiological space_ S _and for_ $*** \in \{mon, cur, pcur\}$,

(a) $Card(\mathbf{Wff}(SAR_S^{***}))$ _is nondenumerably infinite;_

(b) $Card(\mathbf{Wff}(EAR_S^{***}))$ _is nondenumerably infinite._

REMARK 2. From Proposition 2 it follows that SAR_S^{***} and EAR_S^{***} ($*** \in \{mon, cur, pcur\}$) are _not_ arithmetics of SEMIOLOGICAL MATHEMATICS, but of TOPOLOGICAL MATHEMATICS.

In order to transport the arithmetics SAR_S^{***} and EAR_S^{***} back to SEMIOLOGICAL MATHEMATICS, we need to introduce the so-called semicarrier and reversal semicarrier arithmetics in the following subsections.

§4.15. In this section we define and discuss semicarrier-precursive, semicarrier-recursive, and semicarrier-partial-recursive functions and wffs, and then use these to define semicarrier-precursive, semicarrier-recursive, and semicarrier-partial-recursive arithmetics.

We first need the following definitions:

DEFINITION 1. Let S be an atomic nonzeroed semiological space.

(a) If ηSE is a semicarrier such that $\eta : S \longrightarrow E$ is a total surjective function (which need not be injective) and such that η maps the root of space S to $\#_E$ and maps nonroots of space S to nonroots of space E, then we call ηSE a <u>Peano semicarrier space</u> or briefly a <u>Peano semicarrier</u>.

(b) If θES is a semicarrier space such that $\theta : E \longrightarrow S$ is a total injective function, then we call θES a <u>reversal Peano semicarrier space</u> or briefly a <u>reversal Peano semicarrier</u>.

The following proposition is proved in a straightforward manner (see Pogorzelski and Ryan [1982, p. I.2.1.1, Definition 1; p. I.4.1.1, Definition 1; and p. I.2.3.23, Definition 7]):

PROPOSITION 1. <u>Every semiological space S has a denumerably infinite number of Peano subspaces. Moreover, if $\#_S$ is the root of space S, then the number of Peano subspaces of S that have $\#_S$ as their root is equal to the dimension of space S</u> (<u>less</u> 1, <u>in the event that space S is zeroed</u>).

REMARK. We recall that Schwenkel [1965, §2, first paragraph] proves that in the case of a word arithmetic over an infinite alphabet every total function is primitive recursive. Inasmuch as word arithmetics are noncommutative, this result, in the terminology of semiological mathematics, asserts that for every infinite-dimensional (nonzeroed) noncommutative semiological space S we have $\mathbf{F}_S^{mon} = \mathbf{F}_S^{total}$.

However, we note that it is an open question whether Schwenkel's result holds in the case where the semiological space S is commutative. The following definition is needed only for use in the event Schwenkel's result does <u>not</u> hold in the case where space S is commutative. If Schwenkel's result <u>does</u> hold in the case where space S is commutative, then throughout this section we could simply replace $\mathbf{F}_S^{total}$ with $\mathbf{F}_S^{mon}$ or $\mathbf{F}_S^{cur}$, as appropriate, replace $\mathbf{Wff}_S^{total}$ with $\mathbf{Wff}(\mathrm{AR}_S^{mon})$ or $\mathbf{Wff}(\mathrm{AR}_S^{cur})$ ([1]), as appropriate, replace $\mathbf{F}_S^{partial}$ with $\mathbf{F}_S^{pcur}$, and replace $\mathbf{Wff}_S^{partial}$ with $\mathbf{Wff}(\mathrm{AR}_S^{pcur})$.

On the other hand, if Schwenkel's result does <u>not</u> hold in the case where space S is commutative, then, if, e.g., $\mathbf{F}_S^{total}$ were replaced with $\mathbf{F}_S^{mon}$ or $\mathbf{F}_S^{cur}$, as appropriate, throughout this section, the proof of the assertion in Proposition 2 that

---

([1]) For the remainder of this section, for any $*** \in \{mon, pre, cur, rec, pcur, parec\}$, if Σ denotes a suitable semiological space (which may be a magmatical semiological space) or semicarrier space, then we use AR_Σ^{***} to denote either $\mathrm{SAR}_\Sigma^{***}$ or $\mathrm{EAR}_\Sigma^{***}$.

$$\text{for each } g \in \mathbf{F}_E^{\text{pre}} \text{ we have } A_g \neq \emptyset$$

would no longer hold, and, similarly, if, e.g., $\mathbf{Wff}_S^{\text{total}}$ were replaced with $\mathbf{Wff}(\text{AR}_S^{\text{mon}})$ or $\mathbf{Wff}(\text{AR}_S^{\text{cur}})$, as appropriate, throughout this subsection, then the proof of the assertion in Proposition 9 that

$$\text{for each } \mathfrak{u} \in \mathbf{Wff}(\text{AR}_E^{\text{pre}}) \text{ we have } B_{\mathfrak{u}} \neq \emptyset$$

would no longer hold.

DEFINITION 1α. Let S be an atomic nonzeroed infinite-dimensional semiological space.

(i) We define $\mathbf{Wff}_S^{\text{total}}$ to be the set of all wffs obtained by replacing any of all functions in wffs of $\mathbf{Wff}(\text{AR}_S^{\text{cur}})$ by functions in $\mathbf{F}_S^{\text{total}}$.

(ii) We define $\mathbf{Wff}_S^{\text{partial}}$ to be the set of all wffs obtained by replacing any or all functions in wffs of $\mathbf{Wff}(\text{AR}_S^{\text{pcur}})$ by functions in $\mathbf{F}_S^{\text{partial}}$.

DEFINITION 2. Let S be an atomic infinite-dimensional nonzeroed semiological space and either

let ∘∘∘ = total and *** ∈ {pre,rec}

or

let ∘∘∘ = partial and *** = pcur.

Then

(a) For any Peano semicarrier space ηSE, we extend the mapping $\eta : S \longrightarrow E$ to a partial mapping

$$\bar{\eta}: \mathbf{Wff}_S^{\circ\circ\circ} \longrightarrow \mathbf{Wff}(\mathrm{AR}_E^{***})$$

by defining $\bar{\eta}(\mathfrak{w})$ for each $\mathfrak{w} \in \mathbf{Wff}_S^{\circ\circ\circ}$ such that for each function $\mathbf{f}$ occurring in $\mathfrak{w}$ we have $\eta[\mathbf{f}] \in \mathbf{F}_E^{***}$, and in that case we define $\bar{\eta}(\mathfrak{w})$ to be the formula obtained by (i) replacing each occurrence in $\mathfrak{w}$ of a constant $\mathbf{c}$ of S by $\eta(\mathbf{c})$, and (ii) replacing each occurrence in $\mathfrak{w}$ of a function $\mathbf{f}$ by $\eta[\mathbf{f}]$ (which is in $\mathbf{F}_E^{***}$).

We then call $\bar{\eta}$ the *canonical* ***-*extension of* η *to wffs* (where *** is written out as *precursive*, *recursive*, or *partial recursive*, as appropriate).

(b) For any reversal Peano semicarrier space θES, we extend the mapping $\theta : E \longrightarrow S$ to a total mapping

$$\bar{\theta}: \mathbf{Wff}(\mathrm{AR}_E^{***}) \longrightarrow \mathbf{Wff}_S^{\circ\circ\circ}$$

by defining, for each $\mathfrak{u} \in \mathbf{Wff}(\mathrm{AR}_E^{***})$, the value $\bar{\theta}(\mathfrak{u})$ to be the formula obtained by (i) replacing each occurrence in $\mathfrak{u}$ of a constant $\mathbf{c}$ of E by $\theta(\mathbf{c})$, and (ii) replacing each occurrence in $\mathfrak{u}$ of a function $\mathbf{f}$ by $\theta[\mathbf{f}]$ (which is in $\mathbf{F}_S$).

We then call $\bar{\theta}$ the *canonical* ***-*extension of* θ *to wffs* (where *** is written out as *precursive*, *recursive*, or *partial recursive*, as appropriate).

PROPOSITION 1α. *Given the conditions and notation of* Definition 2, *we have the following results:*

$$\text{(i)}\quad \overline{\eta}([\mathbf{t} = u]) = [\overline{\eta}(\mathbf{t}) = \overline{\eta}(u)],$$

where $\mathbf{t}$ *and* u *are terms in any wffs in* $\mathrm{Wff}_S^{\circ\circ\circ}$, *and, in the event that the set* $\mathrm{Wff}_S^{\circ\circ\circ}$ *contains wffs that are not equations,*

$$\text{(ii)}\quad \overline{\eta}(\neg\mathfrak{w}) = \neg\overline{\eta}(\mathfrak{w}),$$

$$\text{(iii)}\quad \overline{\eta}([\mathfrak{w} \wedge \mathfrak{w}']) = [\overline{\eta}(\mathfrak{w}) \wedge \overline{\eta}(\mathfrak{w}')],$$

$$\text{(iv)}\quad \overline{\eta}([\mathfrak{w} \vee \mathfrak{w}']) = [\overline{\eta}(\mathfrak{w}) \vee \overline{\eta}(\mathfrak{w}')],$$

$$\text{(v)}\quad \overline{\eta}([\mathfrak{w} \Rightarrow \mathfrak{w}']) = [\overline{\eta}(\mathfrak{w}) \Rightarrow \overline{\eta}(\mathfrak{w}')],$$

$$\text{(vi)}\quad \overline{\eta}([\mathfrak{w} \Leftrightarrow \mathfrak{w}']) = [\overline{\eta}(\mathfrak{w}) \Leftrightarrow \overline{\eta}(\mathfrak{w}')],$$

where $\mathfrak{w}$ *and* $\mathfrak{w}'$ *are wffs of* $\mathrm{Wff}_S^{\circ\circ\circ}$.

Proof. The proof is straightforward and is by induction on the number of symbols in terms $\mathbf{t}$ and u (in the case of (i)), or the number of symbols in wffs $\mathfrak{w}$ and $\mathfrak{w}'$ (in the case of (ii)-(vi)).

PROPOSITION 1β. *Given the conditions and notation of* Definition 2, *we have the following results:*

$$\text{(i)}\quad \overline{\theta}([\mathbf{t} = u]) = [\overline{\theta}(\mathbf{t}) = \overline{\theta}(u)],$$

where $\mathbf{t}$ *and* u *are terms of* AR_E^{***}, *and, in the event that the arithmetic* AR_E^{***} *is sentential,*

$$\text{(ii)}\quad \overline{\theta}(\neg\mathfrak{w}) = \neg\overline{\theta}(\mathfrak{w}),$$

(iii) $\overline{\theta}([\mathfrak{w} \wedge \mathfrak{w}']) = [\overline{\theta}(\mathfrak{w}) \wedge \overline{\theta}(\mathfrak{w}')]$,

(iv) $\overline{\theta}([\mathfrak{w} \vee \mathfrak{w}']) = [\overline{\theta}(\mathfrak{w}) \vee \overline{\theta}(\mathfrak{w}')]$,

(v) $\overline{\theta}([\mathfrak{w} \Rightarrow \mathfrak{w}']) = [\overline{\theta}(\mathfrak{w}) \Rightarrow \overline{\theta}(\mathfrak{w}')]$,

(vi) $\overline{\theta}([\mathfrak{w} \Leftrightarrow \mathfrak{w}']) = [\overline{\theta}(\mathfrak{w}) \Leftrightarrow \overline{\theta}(\mathfrak{w}')]$,

where $\mathfrak{w}$ and $\mathfrak{w}'$ are wffs in $\mathrm{Wff}(\mathrm{AR}_E^{***})$.

Proof. The proof is straightforward and is by induction on the number of symbols in terms **t** and u (in the case of (i)), or the number of symbols in wffs $\mathfrak{w}$ and $\mathfrak{w}'$ (in the case of (ii)-(vi)).

For the remainder of this section we let S be an atomic infinite-dimensional nonzeroed semiological space, let ηSE be a Peano semicarrier space, and for

$$*** \in \{\text{precursive, recursive, partial recursive}\}$$

let $\overline{\eta}$ be the canonical ***-extension of η to wffs.

DEFINITION 3. Let S and ηSE be as given in the paragraph preceding this definition.

(a) For any $f \in \mathrm{F}_S^{\mathrm{total}}$, we call f an ηSE-precursive S-function or semicarrier-precursive S-function with respect to ηSE if and only if $\eta[f] \in \mathrm{F}_E^{\mathrm{pre}}$.

We let $\mathfrak{F}_{\eta SE}^{\mathrm{scar\text{-}pre}}$ denote the set of all semicarrier-precursive S-functions with respect to ηSE.

For any $g \in \mathrm{F}_E^{\mathrm{pre}}$ and any $f \in \mathrm{F}_S^{\mathrm{total}}$, if $\eta[f] = g$, then we call f a <u>precursive repetition of</u> g. In turn, for each $g \in \mathbf{F}_E^{\text{pre}}$, we call the set $\{f \mid f \in \mathbf{F}_S^{\text{total}}$ and $\eta[f] = g\}$ the <u>set of precursive repetitions of</u> g.

(b) For any $f \in \mathbf{F}_S^{\text{total}}$, we call f an ηSE-<u>recursive</u> S-<u>function</u> or a <u>semicarrier-recursive</u> S-<u>function with respect to</u> ηSE if and only if $\eta[f] \in \mathbf{F}_E^{\text{rec}}$.

We let $\mathfrak{F}_{\eta SE}^{\text{scar-rec}}$ denote the set of all semicarrier-recursive S-functions with respect to ηSE.

For any $g \in \mathbf{F}_E^{\text{rec}}$ and any $f \in \mathbf{F}_S^{\text{total}}$, if $\eta[f] = g$, then we call f a <u>recursive repetition of</u> g. In turn, for each $g \in \mathbf{F}_E^{\text{rec}}$, we call the set

$$\{f \mid f \in \mathbf{F}_S^{\text{total}} \text{ and } \eta[f] = g\}$$

the <u>set of recursive repetitions of</u> g.

(c) For any $f \in \mathbf{F}_S^{\text{partial}}$, we call f an ηSE-<u>partial-recursive</u> S-<u>function</u> or a <u>semicarrier-partial-recursive</u> S-<u>function with respect to</u> ηSE if and only if $\eta[f] \in \mathbf{F}_E^{\text{parec}}$.

We let $\mathfrak{F}_{\eta SE}^{\text{scar-parec}}$ denote the set of all semicarrier-partial-recursive S-functions with respect to ηSE.

For any $g \in \mathbf{F}_E^{\text{parec}}$ and any $f \in \mathbf{F}_S^{\text{partial}}$, if $\eta[f] = g$, then we call f a <u>partial-recursive-repetition of</u> g. In turn, for each $g \in \mathbf{F}_E^{\text{parec}}$, we call the set

$$\{f \mid f \in \mathbf{F}_S^{\text{partial}} \text{ and } \eta[f] = g\}$$

the <u>set of partial recursive repetitions of</u> g.

PROPOSITION 2. Let S and ηSE be as given in the paragraph preceding Definition 3, and define a family $\mathscr{F} = (A_g)_{g \in \mathbf{F}_E^{\text{pre}}}$ by defining for each $g \in \mathbf{F}_E^{\text{pre}}$:

$$A_g = \{f \mid f \in \mathbf{F}_S^{\text{total}} \text{ and } \eta[f] = g\}.$$

(We see that each such set A_g is the set of precursive repetitions of g.) Then the family $\mathscr{F}$ is a partition of the set of functions in $\mathbf{F}_S^{\text{total}}$ that are mapped by η to functions in $\mathbf{F}_E^{\text{pre}}$, so that $\mathscr{F}$ is a family of equivalence classes of precursive repetitions. Moreover, $\mathfrak{F}_{\eta SE}^{\text{scar-pre}}$ is the set of precursive repetitions of functions in $\mathbf{F}_E^{\text{pre}}$. Lastly, for each $g \in \mathbf{F}_E^{\text{pre}}$ we have $A_g \neq \emptyset$.

Proof. First, from the definition of $\mathscr{F}$ we see that each A_g ($g \in \mathbf{F}_E^{\text{pre}}$) is a set of functions in $\mathbf{F}_S^{\text{total}}$ that are mapped by η to g.

In turn, for some $f \in \mathbf{F}_S^{\text{total}}$ assume that η maps f to some function g in $\mathbf{F}_E^{\text{pre}}$; i.e., assume that $\eta[f] = g$. Then by the definition of $\mathscr{F}$ we have $f \in A_g$.

Next, for some g and g' in $\mathbf{F}_E^{\text{pre}}$ and some $f \in \mathbf{F}_S^{\text{total}}$, assume that $f \in A_g$ and $f \in A_{g'}$. Then by the definition of $\mathscr{F}$ we have $\eta[f] = g$ and $\eta[f] = g'$, so that $g = g'$, which yields $A_g = A_{g'}$, which establishes the first assertion of the proposition.

By part (a) of Definition 3,

$$\mathfrak{F}^{\text{scar-pre}}_{\eta SE} = \{f \mid f \in \mathbf{F}^{\text{total}}_{S} \text{ and } (\exists g)(g \in \mathbf{F}^{\text{pre}}_{E} \text{ and } \eta[f] = g)\}$$

$$= \{f \mid f \in \mathbf{F}^{\text{total}}_{S} \text{ and } f \text{ is a precursive repetition of some } g \in \mathbf{F}^{\text{pre}}_{E}\}.$$

That is, $\mathfrak{F}^{\text{scar-pre}}_{\eta SE}$ is the set of precursive repetitions of functions in $\mathbf{F}^{\text{pre}}_{E}$, which proves the second assertion of the proposition.

To establish the final assertion of the proposition, let $g \in \mathbf{F}^{\text{pre}}_{E}$ be arbitrary and for some $n \geq 1$ assume that g is an n-place function. We then construct a function $f \in \mathbf{F}^{\text{total}}_{S}$ such that $\eta[f] = g$:

For any $a_1, a_2, \ldots, a_n, b \in E$, assume that

$$(*) \qquad g(a_1, a_2, \ldots, a_n) = b.$$

Recalling (Definition 1 above) that $\eta: S \longrightarrow E$ is a total surjection, we see that there exist $c_1, c_2, \ldots, c_n, d \in S$ such that $\eta(c_i) = a_i$ $(1 \leq i \leq n)$ and $\eta(d) = b$. (There may be more than one such series $c_1, c_2, \ldots, c_n, d$, but we can select <u>one</u> (in particular, we can select <u>one</u> such d), using Bourbaki's τ-operator, if necessary.) Then for <u>all</u> $c'_1, c'_2, \ldots, c'_n \in S$ such that $\eta(c'_i) = a_i$ $(1 \leq i \leq n)$ we define

$$f(c'_1, c'_2, \ldots, c'_n) = d,$$

or, in other notation,

$$((c'_1, c'_2, \ldots, c'_n), d) \in f,$$

in which case (Pogorzelski and Ryan [**1982**, Chapter 0, §1, no. 3])

$$((\eta(c_1'),\eta(c_2'),\dots,\eta(c_n')),\eta(d)) \in \eta[f],$$

that is, $((a_1,a_2,\dots,a_n),b) \in \eta[f]$, or, in other notation,

$$(**) \qquad \eta[f](a_1,a_2,\dots,a_n) = b.$$

We see that we have, in fact, defined an n-place S-function f.

Now let c_1, c_2, ..., $c_n \in S$. Then from $\eta: S \longrightarrow E$ being total it follows that $\eta(c_1)$, $\eta(c_2)$, ..., $\eta(c_n)$ are all defined and belong to E, so that $g(\eta(c_1),\eta(c_2),\dots,\eta(c_n))$ is defined and for some $b \in E$ we have

$$g(\eta(c_1),\eta(c_2),\dots,\eta(c_n)) = b.$$

From $b \in E$ and $\eta: S \longrightarrow E$ being surjective it follows that there exists a $d \in S$ such that $\eta(d) = b$, and by the definition of f we have $f(c_1,c_2,\dots,c_n) = d$.

Thus we have proved that f is a total n-place S-function so that $f \in \mathbf{F}_S^{\text{total}}$. Finally, from (*) and (**) we have $\eta[f] = g$, so that by the definition of the family $\mathcal{F}$ in the statement of the proposition we have $f \in A_g$, so that $A_g \neq \emptyset$.

PROPOSITION 3. <u>Let</u> S <u>and</u> ηSE <u>be as given in the paragraph preceding</u> Definition 3, <u>and define a family</u> $\mathcal{F} = (A_g)_{g \in \mathbf{F}_E^{\text{rec}}}$ <u>by defining for each</u> $g \in \mathbf{F}_E^{\text{rec}}$:

$$A_g = \{f \mid f \in \mathbf{F}_S^{\text{total}} \text{ and } \eta[f] = g\}.$$

(<u>We see that each such set A_g is the set of recursive repetitions of g.) Then the family $\mathscr{F}$ is a partition of the set of functions in $\mathbf{F}_S^{\text{total}}$ that are mapped by η to functions in $\mathbf{F}_E^{\text{rec}}$, so that $\mathscr{F}$ is a family of equivalence classes of recursive repetitions. Moreover, $\mathfrak{F}_{\eta SE}^{\text{scar-rec}}$ is the set of recursive repetitions of functions in $\mathbf{F}_E^{\text{rec}}$. Lastly, for each $g \in \mathbf{F}_E^{\text{rec}}$ we have $A_g \neq \emptyset$.</u>

<u>Proof</u>. One uses the proof of Proposition 2 with $\mathbf{F}_E^{\text{pre}}$ replaced by $\mathbf{F}_E^{\text{rec}}$ and with the reference to part (a) of Definition 3 replaced by a reference to part (b) of Definition 3.

PROPOSITION 4. <u>Let S and ηSE be as given in the paragraph preceding</u> Definition 3, <u>and define a family</u> $\mathscr{F} = (A_g)_{g \in \mathbf{F}_E^{\text{parec}}}$ <u>by defining for each $g \in \mathbf{F}_E^{\text{parec}}$:</u>

$$A_g = \{f \mid f \in \mathbf{F}_S^{\text{partial}} \text{ and } \eta[f] = g\}.$$

(<u>We see that each such set A_g is the set of partial recursive repetitions of g.) Then the family $\mathscr{F}$ is a partition of the set of functions in $\mathbf{F}_S^{\text{partial}}$ that are mapped by η to functions in $\mathbf{F}_E^{\text{parec}}$, so that $\mathscr{F}$ is a family of equivalence classes of partial recursive repetitions. Moreover, $\mathfrak{F}_{\eta SE}^{\text{scar-parec}}$ is the set of partial recursive repetitions of functions in $\mathbf{F}_E^{\text{parec}}$. Lastly, for each $g \in \mathbf{F}_E^{\text{parec}}$ we have $A_g \neq \emptyset$.</u>

<u>Proof</u>. The proof is along the lines of the proof of

Proposition 2, with some changes made to accommodate the fact that $F_S^{partial}$ and F_E^{parec} both contain functions that are not total.

PROPOSITION 5. Let S and ηSE be as given in the paragraph preceding Definition 3, let $*** \in \{mon, cur, pcur\}$ ([2]), and let the family $\mathcal{F} = (A_g)_{g \in F_E^{***}}$ be as defined in Propositions 2-4. Then

$$\mathfrak{F}_{\eta SE}^{scar-***} = \bigcup A_g,$$

where g ranges over F_E^{***}.

Proof. The proof is straightforward, using Definition 3 and Propositions 2-4.

Before giving the following definition, we point out that from S being a semiological space (and hence being a denumerable set) it follows that the elements of S can be enumerated, say a_0, a_1, a_2, ... (with $a_0 = \#_S$):

DEFINITION 4. Let S and ηSE be as given in the paragraph preceding Definition 3, let a_0, a_1, a_2, ... (with $a_0 = \#_S$) be an enumeration of the elements of S, and let a total function

$$\chi : F_E^{pre} \longrightarrow F_S^{total}$$

be given such that for each $g \in F_E^{pre}$ we have $\chi(g) \in A_g$ (that

---

([2]) Recall footnote 1 of the present section.

is, for each $g \in F_E^{pre}$, the value $\chi(g)$ is a function in F_S^{total} that is mapped by η to g, i.e., we have $\eta[\chi(g)] = g$), where the following condition is satisfied:

(*) for each $k \geq 0$, there exists a $g \in F_E^{pre}$ such that $\chi(g)(a_k) = a_{k+1}$ (3).

(One sees that such a function χ is along the lines of a choice function, so that it can be constructed using the Hilbert ε-operator, if desired.) We then let

$$F_{\eta SE}^{scar-\chi-pre} = \{\chi(g) \mid g \in F_E^{pre}\}.$$

In turn, we define a set A_{SE}, whose elements are denoted as μ_0, μ_1, μ_2, ..., to be a denumerably infinite subset of

$$F_{\eta SE}^{scar-\chi-pre}$$

such that for each $k \geq 0$ we have $\mu_k(a_k) = a_{k+1}$. We call A_{SE} the <u>semicarrier</u>-χ-<u>successor</u> <u>set</u> (or briefly a <u>semicarrier</u> <u>successor</u> <u>set</u>) and call the functions μ_0, μ_1, μ_2, ... <u>semicarrier</u>-χ-<u>successor</u> <u>functions</u> (or briefly <u>semicarrier</u> <u>successor</u> <u>functions</u>). We note that the set A_{SE}, as well as the function χ, can be constructed using the Hilbert ε-operator, if necessary.

---

(3) In Proposition 5α we prove that for each $k \geq 0$ there does exist a function $\mu_k \in F_S^{total}$ such that $\eta[\mu_k] \in F_E^{pre}$ and such that $\mu_k(a_k) = a_{k+1}$, so that (with $\chi(\eta[\mu_k]) = \mu_k$ (in the present definition)) condition (*) can be shown to hold.

PROPOSITION 5α. *Let S and η^{SE} be as given in the paragraph preceding* Definition 3 *and let $a_0, a_1, a_2, \ldots$ (with $a_0 = \#_S$) be an enumeration of the elements of S. Then for each $k \geq 0$ there exists a function (not necessarily unique) $\mu_k \in \mathbf{F}_S^{\text{total}}$ such that $\mu_k(a_k) = a_{k+1}$ and such that $\eta[\mu_k] \in \mathbf{F}_E^{\text{pre}}$. In other words a successor set A_{SE} exists.*

Proof. For each $k \geq 0$, consider the constant S-function μ_k that satisfies:

$$\mu_k(x) = a_{k+1} \qquad (x \in S).$$

We see that $\eta[\mu_k]$ is an E-function that satisfies

$$\eta[\mu_k](x) = \eta(a_{k+1}) \qquad (x \in E),$$

in which case we clearly have $\eta[\mu_k] \in \mathbf{F}_E^{\text{pre}}$.

PROPOSITION 5β. *Assume the conditions and notation of* Definition 4. *Then every $a_n \in S$ different from $\#_S$ (in which case $n > 1$) can be expressed as the resolution*

$$a_n = \mu_{n-1} \circ \cdots \circ \mu_2 \circ \mu_1(\#_S),$$

where $\mu_0, \mu_1, \ldots, \mu_{n-1} \in A_{SE}$.

Proof. Recalling that we are denoting an enumeration of the elements of S as $a_0, a_1, a_2, \ldots$ (with $a_0 = \#_S$), we prove the proposition for arbitrary $a_k \in S$ ($k \geq 1$), using induction on k.

We first consider the case of a_1. By Definition 4, we have $\mu_0(\#_S) = \mu_0(a_0) = a_1$, and the proposition holds for $k = 1$.

Assume as induction hypothesis that for some $k \geq 1$ we have

$$a_k = \mu_{k-1} \circ \cdots \circ \mu_1 \circ \mu_0(\#_S).$$

Then by Definition 4 we have $\mu_k(a_k) = a_{k+1}$, which with the induction hypothesis yields

$$a_{k+1} = \mu_k(\mu_{k-1} \circ \cdots \circ \mu_1 \circ \mu_0(\#_S)$$

$$= \mu_k \circ \mu_{k-1} \circ \cdots \circ \mu_1 \circ \mu_0(\#_S).$$

Thus the proposition holds for $k + 1$, and so holds generally.

DEFINITION 5. Let S and ηSE be as given in the paragraph preceding Definition 3, and let a total function

$$\chi : \mathbf{F}_E^{\text{rec}} \longrightarrow \mathbf{F}_S^{\text{total}}$$

be given such that for each $g \in \mathbf{F}_E^{\text{rec}}$ we have $\chi(g) \in A_g$ (that is, for each $g \in \mathbf{F}_E^{\text{rec}}$, $\chi(g)$ is a function in $\mathbf{F}_S^{\text{total}}$ that is mapped by η to g, i.e., we have $\eta[\chi(g)] = g$). We then let

$$\mathbf{F}_{\eta SE}^{\text{scar-}\chi\text{-rec}} = \{\chi(g) \,|\, g \in \mathbf{F}_E^{\text{rec}}\}.$$

DEFINITION 6. Let S and ηSE be as given in the paragraph preceding Definition 3, and let a total function

$$\chi : \mathbf{F}_E^{\text{parec}} \longrightarrow \mathbf{F}_S^{\text{partial}}$$

be given such that for each $g \in \mathbf{F}_E^{\text{parec}}$ we have $\chi(g) \in A_g$ (that is, for each $g \in \mathbf{F}_E^{\text{parec}}$, $\chi(g)$ is a function in $\mathbf{F}_S^{\text{partial}}$ that is mapped by η to g, i.e., we have $\eta[\chi(g)] = g$). We then let

$$F_{\eta SE}^{\text{scar-}\chi\text{-parec}} = \{\chi(g) \mid g \in F_E^{\text{parec}}\}.$$

PROPOSITION 6. Let S and ηSE be as given in the paragraph preceding Definition 3, and let a total function

$$\chi: F_E^{\text{pre}} \longrightarrow F_S^{\text{total}}$$

and the set $F_{\eta SE}^{\text{scar-}\chi\text{-pre}}$ satisfy the conditions given in Definition 4. Then $F_{\eta SE}^{\text{scar-}\chi\text{-pre}}$ is a set of precursive repetitions of functions in F_E^{pre}; moreover, for each function $g \in F_E^{\text{pre}}$ there is a unique function $f \in F_{\eta SE}^{\text{scar-}\chi\text{-pre}}$ such that $\eta[f] = g$. In other words, each function in F_E^{pre} has exactly one precursive repetition in $F_{\eta SE}^{\text{scar-}\chi\text{-pre}}$. Then we can say that $\chi: F_E^{\text{pre}} \longrightarrow F_{\eta SE}^{\text{scar-}\chi\text{-pre}}$ is a bijection, so that from F_E^{pre} being denumerably infinite it follows that $F_{\eta SE}^{\text{scar-}\chi\text{-pre}}$ is also denumerably infinite.

Proof. The proof is straightforward, using part (a) of Definition 3, Definition 4, and Proposition 2.

PROPOSITION 7. Let S and ηSE be as given in the paragraph preceding Definition 3, and let a total function

$$\chi: F_E^{\text{rec}} \longrightarrow F_S^{\text{total}}$$

and the set $F_{\eta SE}^{\text{scar-}\chi\text{-rec}}$ satisfy the conditions given in Definition 5. Then $F_{\eta SE}^{\text{scar-}\chi\text{-rec}}$ is a set of recursive repetitions of functions in F_E^{rec}; moreover, for each function $g \in F_E^{\text{rec}}$ there is a unique function $f \in F_{\eta SE}^{\text{scar-}\chi\text{-rec}}$ such that $\eta[f] = g$. In other words, each function in F_E^{rec} has exactly one recursive repetition in $F_{\eta SE}^{scar-\chi-rec}$. Then we can say that $\chi: F_E^{rec} \longrightarrow F_{\eta SE}^{scar-\chi-rec}$ is a bijection, so that from F_E^{rec} being denumerably infinite it follows that $F_{\eta SE}^{scar-\chi-rec}$ is also denumerably infinite.

Proof. The proof is straightforward, using part (b) of Definition 3, Definition 5, and Proposition 3.

PROPOSITION 8. Let S and ηSE be as given in the paragraph preceding Definition 3, and let a total function

$$\chi: F_E^{parec} \longrightarrow F_S^{partial}$$

and the set $F_{\eta SE}^{scar-\chi-parec}$ satisfy the conditions given in Definition 6. Then $F_{\eta SE}^{scar-\chi-parec}$ is a set of partial recursive repetitions of functions in F_E^{parec}; moreover, for each function $g \in F_E^{parec}$ there is a unique function

$$f \in F_{\eta SE}^{scar-\chi-parec}$$

such that $\eta[f] = g$. In other words, each function in F_E^{parec} has exactly one partial recursive repetition in

$$F_{\eta SE}^{scar-\chi-parec}.$$

Then we can say that $\chi: F_E^{parec} \longrightarrow F_{\eta SE}^{scar-\chi-parec}$ is a bijection, so that from F_E^{parec} being denumerably infinite it follows that $F_{\eta SE}^{scar-\chi-parec}$ is also denumerably infinite.

Proof. The proof is straightforward, using part (c) of Definition 3, Definition 6, and Proposition 4.

DEFINITION 7. Let S, ηSE, and $\bar{\eta}$ be as given in the paragraph preceding Definition 3.

(a) For any $\mathfrak{w} \in \mathrm{Wff}_S^{\mathrm{total}}$, we call $\mathfrak{w}$ an ηSE-precursive S-wff or a semicarrier-precursive S-wff with respect to ηSE if and only if $\bar{\eta}(\mathfrak{w}) \in \mathrm{Wff}(\mathrm{AR}_E^{\mathrm{pre}})$.

We let $\mathfrak{W}_{\eta SE}^{\mathrm{scar\text{-}pre}}$ denote the set of all semicarrier-precursive S-wffs with respect to ηSE.

For any $\mathfrak{u} \in \mathrm{Wff}(\mathrm{AR}_E^{\mathrm{pre}})$ and any $\mathfrak{w} \in \mathrm{Wff}_S^{\mathrm{total}}$, if $\bar{\eta}(\mathfrak{w}) = \mathfrak{u}$, then we call $\mathfrak{w}$ a precursive repetition of $\mathfrak{u}$. In turn, for each $\mathfrak{u} \in \mathrm{Wff}(\mathrm{AR}_E^{\mathrm{pre}})$, we call the set

$$\{\mathfrak{w} \mid \mathfrak{w} \in \mathrm{Wff}_S^{\mathrm{total}} \text{ and } \bar{\eta}(\mathfrak{w}) = \mathfrak{u}\}$$

the set of precursive repetitions of $\mathfrak{u}$.

(b) For any $\mathfrak{w} \in \mathrm{Wff}_S^{\mathrm{total}}$, we call $\mathfrak{w}$ an ηSE-recursive S-wff or a semicarrier-recursive S-wff with respect to ηSE if and only if $\bar{\eta}(\mathfrak{w}) \in \mathrm{Wff}(\mathrm{AR}_E^{\mathrm{rec}})$.

We let $\mathfrak{W}_{\eta SE}^{\mathrm{scar\text{-}rec}}$ denote the set of all semicarrier-recursive S-wffs with respect to ηSE.

For any $\mathfrak{u} \in \mathrm{Wff}(\mathrm{AR}_E^{\mathrm{rec}})$ and any $\mathfrak{w} \in \mathrm{Wff}_S^{\mathrm{total}}$, if $\bar{\eta}(\mathfrak{w}) = \mathfrak{u}$, then we call $\mathfrak{w}$ a recursive repetition of $\mathfrak{u}$. In turn, for each $\mathfrak{u} \in \mathrm{Wff}(\mathrm{AR}_E^{\mathrm{rec}})$, we call the set

$$\{\mathfrak{w} \mid \mathfrak{w} \in \mathrm{Wff}_S^{\mathrm{total}} \text{ and } \bar{\eta}(\mathfrak{w}) = \mathfrak{u}\}$$

the set of recursive repetitions of $\mathfrak{u}$.

(c) For any $\mathfrak{w} \in \mathrm{Wff}_S^{\mathrm{partial}}$, we call $\mathfrak{w}$ an ηSE-partial-recursive S-wff or a semicarrier-partial-recursive

S-wff with respect to ηSE if and only if $\bar{\eta}(\mathfrak{w}) \in \mathrm{Wff}(\mathrm{AR}_E^{\mathrm{parec}})$.

We let $\mathfrak{W}_{\eta SE}^{\mathrm{scar\text{-}parec}}$ denote the set of all semicarrier-recursive S-wffs with respect to ηSE.

For any $\mathfrak{u} \in \mathrm{Wff}(\mathrm{AR}_E^{\mathrm{parec}})$ and any $\mathfrak{w} \in \mathrm{Wff}_S^{\mathrm{partial}}$, if $\bar{\eta}(\mathfrak{w}) = \mathfrak{u}$, then we call $\mathfrak{w}$ a partial recursive repetition of $\mathfrak{u}$. In turn, for each $\mathfrak{u} \in \mathrm{Wff}(\mathrm{AR}_E^{\mathrm{parec}})$, we call the set

$$\{\mathfrak{w} \mid \mathfrak{w} \in \mathrm{Wff}_S^{\mathrm{partial}} \text{ and } \bar{\eta}(\mathfrak{w}) = \mathfrak{u}\}$$

the set of partial recursive repetitions of $\mathfrak{u}$.

PROPOSITION 9. Let S, ηSE, and $\bar{\eta}$ be as given in the paragraph preceding Definition 3, and define a family

$$\mathcal{W} = (B_{\mathfrak{u}})_{\mathfrak{u} \in \mathrm{Wff}(\mathrm{AR}_E^{\mathrm{pre}})}$$

by defining for each $\mathfrak{u} \in \mathrm{Wff}(\mathrm{AR}_E^{\mathrm{pre}})$:

$$B_{\mathfrak{u}} = \{\mathfrak{w} \mid \mathfrak{w} \in \mathrm{Wff}_S^{\mathrm{total}} \text{ and } \bar{\eta}(\mathfrak{w}) = \mathfrak{u}\}.$$

(We see that each such set $B_{\mathfrak{u}}$ is the set of precursive repetitions of $\mathfrak{u}$.) Then the family $\mathcal{W}$ is a partition of the set of wffs in $\mathrm{Wff}_S^{\mathrm{total}}$ that are mapped by $\bar{\eta}$ to wffs in $\mathrm{Wff}(\mathrm{AR}_E^{\mathrm{pre}})$, so that $\mathcal{W}$ is a family of equivalence classes of precursive repetitions. Moreover, $\mathfrak{W}_{\eta SE}^{\mathrm{scar\text{-}pre}}$ is the set of precursive repetitions of the wffs in $\mathrm{Wff}(\mathrm{AR}_E^{\mathrm{pre}})$. Lastly, for each $\mathfrak{u} \in \mathrm{Wff}(\mathrm{AR}_E^{\mathrm{pre}})$ we have $B_{\mathfrak{u}} \neq \emptyset$.

Proof. Proof is straightforward, using the definition of the set $\mathcal{W}$ and part (a) of Definition 7.

PROPOSITION 10. Let S, ηSE, and $\overline{\eta}$ be as given in the paragraph preceding Definition 3, and define a family

$$\mathcal{W} = (B_{\mathfrak{u}})_{\mathfrak{u} \in \mathrm{Wff}(\mathrm{AR}_E^{\mathrm{rec}})}$$

by defining for each $\mathfrak{u} \in \mathrm{Wff}(\mathrm{AR}_E^{\mathrm{rec}})$:

$$B_{\mathfrak{u}} = \{\mathfrak{w} \mid \mathfrak{w} \in \mathrm{Wff}_S^{\mathrm{total}} \text{ and } \overline{\eta}(\mathfrak{w}) = \mathfrak{u}\}.$$

(We see that each such set $B_{\mathfrak{u}}$ is the set of recursive repetitions of $\mathfrak{u}$.) Then the family $\mathcal{W}$ is a partition of the set of wffs in $\mathrm{Wff}_S^{\mathrm{total}}$ that are mapped by $\overline{\eta}$ to wffs in $\mathrm{Wff}(\mathrm{AR}_E^{\mathrm{rec}})$, so that $\mathcal{W}$ is a family of equivalence classes of recursive repetitions. Moreover, $\mathfrak{W}_{\eta SE}^{\mathrm{scar\text{-}rec}}$ is the set of recursive repetitions of the wffs in $\mathrm{Wff}(\mathrm{AR}_E^{\mathrm{rec}})$. Lastly, for each $\mathfrak{u} \in \mathrm{Wff}(\mathrm{AR}_E^{\mathrm{rec}})$ we have $B_{\mathfrak{u}} \neq \emptyset$.

Proof. Proof is straightforward, using the definition of the set $\mathcal{W}$ and part (b) of Definition 7.

PROPOSITION 11. Let S, ηSE, and $\overline{\eta}$ be as given in the paragraph preceding Definition 3, and define a family

$$\mathcal{W} = (B_{\mathfrak{u}})_{\mathfrak{u} \in \mathrm{Wff}(\mathrm{AR}_E^{\mathrm{parec}})}$$

by defining for each $\mathfrak{u} \in \mathrm{Wff}(\mathrm{AR}_E^{\mathrm{parec}})$:

$$B_{\mathfrak{u}} = \{\mathfrak{w} \mid \mathfrak{w} \in \mathrm{Wff}_S^{\mathrm{partial}} \text{ and } \overline{\eta}(\mathfrak{w}) = \mathfrak{u}\}.$$

(We see that each such set $B_{\mathfrak{u}}$ is the set of partial recursive repetitions of $\mathfrak{u}$.) Then the family $\mathcal{W}$ is a partition of the set of wffs in $\mathrm{Wff}_S^{\mathrm{partial}}$ that are mapped by $\bar{\eta}$ to wffs in $\mathrm{Wff}(\mathrm{AR}_E^{\mathrm{parec}})$, so that $\mathcal{V}$ is a family of equivalence classes of partial recursive repetitions. Moreover, $\mathfrak{W}_{\eta SE}^{\mathrm{scar\text{-}parec}}$ is the set of partial recursive repetitions of the wffs in $\mathrm{Wff}(\mathrm{AR}_E^{\mathrm{parec}})$. Lastly, for each $\mathfrak{u} \in \mathrm{Wff}(\mathrm{AR}_E^{\mathrm{parec}})$, we have $B_{\mathfrak{u}} \neq \emptyset$.

Proof. Proof is straightforward, using the definition of the set $\mathcal{V}$ and part (c) of Definition 7.

PROPOSITION 12. Let S, ηSE, and $\bar{\eta}$ be as given in the paragraph preceding Definition 3, let $*** \in \{\mathrm{mon}, \mathrm{cur}, \mathrm{pcur}\}$, and let the family

$$\mathcal{V} = (B_{\mathfrak{u}})_{\mathfrak{u} \in \mathrm{Wff}(\mathrm{AR}_E^{***})}$$

be as defined in Propositions 9, 10, and 11. Then

$$\mathfrak{W}_{\eta SE}^{\mathrm{scar\text{-}***}} = \bigcup B_{\mathfrak{u}},$$

where $\mathfrak{u}$ ranges over $\mathrm{Wff}(\mathrm{AR}_E^{***})$.

Proof. By Definition 7 and Propositions 9, 10, and 11.

PROPOSITION 13. Let S, ηSE, and $\bar{\eta}$ be as given in the paragraph preceding Definition 3, and let a mapping

$$\chi : \mathrm{F}_E^{\mathrm{pre}} \longrightarrow \mathrm{F}_S^{\mathrm{total}}$$

and the set $F_{\eta SE}^{scar-\chi-pre}$ satisfy the conditions given in Definition 4. Then for each wff $\mathfrak{u} \in Wff(AR_E^{pre})$ there exist at most a denumerably infinite number of wffs $\mathfrak{w} \in \mathfrak{W}_{\eta SE}^{scar-pre}$ such that $\overline{\eta}(\mathfrak{w}) = \mathfrak{u}$ and such that all functions occurring in $\mathfrak{w}$ are functions in $F_{\eta SE}^{scar-\chi-pre}$. Thus if we denote as

$$Wff(AR_{\eta SE}^{scar-\chi-pre}) \ (^4)$$

the set of wffs in $\mathfrak{W}_{\eta SE}^{scar-pre}$ all of whose functions are in $F_{\eta SE}^{scar-\chi-pre}$, then the set $Wff(AR_{\eta SE}^{scar-\chi-pre})$ is denumerably infinite.

Proof. Follows from $F_{\eta SE}^{scar-\chi-pre}$ being denumerably infinite (Proposition 6), the terms in S being denumerably infinite, the set of variables being denumerably infinite, and the set of improper symbols being finite, so that the set of wffs all of whose functions are in $F_{\eta SE}^{scar-\chi-pre}$ can be Gödel-numbered and hence is at most denumerably infinite. By the final assertion in Proposition 2 there exists an $f \in F_S^{total}$ such that $\eta[f] = \sigma_E$, and so each of the wffs

$$\#_S = \#_S,\ f(\#_S) = f(\#_S),\ f(f(\#_S)) = f(f(\#_S)),\ \ldots$$

belongs to $Wff(AR_{\eta SE}^{scar-\chi-pre})$, and so the set of wffs all of whose functions are in $F_{\eta SE}^{scar-\chi-pre}$ is exactly denumerably infinite.

---

(4) We note that for $*** \in \{pre, rec, parec\}$, the arithmetics $AR_{\eta SE}^{scar-\chi-***}$ are defined in Definitions 8-10 below.

PROPOSITION 14. Let S, ηSE, and $\overline{\eta}$ be as given in the paragraph preceding Definition 3, and let a mapping

$$\chi : F_E^{rec} \longrightarrow F_S^{total}$$

and the set $F_{\eta SE}^{scar-\chi-rec}$ satisfy the conditions given in Definition 5. Then for each wff $\mathfrak{u} \in Wff(AR_E^{rec})$ there exist at most a denumerably infinite number of wffs $\mathfrak{w} \in \mathfrak{W}_{\eta SE}^{scar-rec}$ such that $\overline{\eta}(\mathfrak{w}) = \mathfrak{u}$ and such that all functions occurring in $\mathfrak{w}$ are functions in $F_{\eta SE}^{scar-\chi-rec}$. Thus if we denote as

$$Wff(AR_{\eta SE}^{scar-\chi-rec})$$

the set of wffs in $\mathfrak{W}_{\eta SE}^{scar-rec}$ all of whose functions are in $F_{\eta SE}^{scar-\chi-rec}$, then the set $Wff(AR_{\eta SE}^{scar-\chi-rec})$ is denumerably infinite.

Proof. Similar to the proof of Proposition 13.

PROPOSITION 15. Let S, ηSE, and $\overline{\eta}$ be as given in the paragraph preceding Definition 3, and let a mapping

$$\chi : F_E^{parec} \longrightarrow F_S^{partial}$$

and the set $F_{\eta SE}^{scar-\chi-parec}$ satisfy the conditions given in Definition 6. Then for each wff $\mathfrak{u} \in Wff(AR_E^{parec})$ there exists at most a denumerably infinite number of wffs

$$\mathfrak{w} \in \mathfrak{W}_{\eta SE}^{scar-parec}$$

such that $\overline{\eta}(\mathfrak{w}) = \mathfrak{u}$ and such that all functions occurring in $\mathfrak{w}$ are functions in $F_{\eta SE}^{scar-\chi-parec}$. Thus if we denote as

$$\mathrm{Wff}(\mathrm{AR}_{\eta SE}^{\mathrm{scar}\text{-}\chi\text{-}\mathrm{parec}})$$

<u>the set of wffs in</u> $\mathfrak{W}_{\eta SE}^{\mathrm{scar}\text{-}\mathrm{parec}}$ <u>all of whose functions are in</u> $\mathrm{F}_{\eta SE}^{\mathrm{scar}\text{-}\chi\text{-}\mathrm{parec}}$, <u>then the set</u> $\mathrm{Wff}(\mathrm{AR}_{\eta SE}^{\mathrm{scar}\text{-}\chi\text{-}\mathrm{parec}})$ <u>is denumerably infinite</u>.

<u>Proof</u>. Similar to the proof of Proposition 13.

DEFINITION 8. Let S, ηSE, and $\overline{\eta}$ be as given in the paragraph preceding Definition 3, and let a function $\chi\colon \mathrm{F}_{E}^{\mathrm{pre}} \longrightarrow \mathrm{F}_{S}^{\mathrm{total}}$ be as given in Proposition 6.

In this definition we simultaneously define the <u>sentential semicarrier</u>-χ-<u>precursive arithmetic</u>, denoted as $\mathrm{SAR}_{\eta SE}^{\mathrm{scar}\text{-}\chi\text{-}\mathrm{pre}}$, or briefly as $\mathrm{SAR}_{\eta SE}^{\chi\text{-}\mathrm{pre}}$, and the <u>equational semicarrier</u>-χ-<u>precursive arithmetic</u>, denoted as $\mathrm{EAR}_{\eta SE}^{\mathrm{scar}\text{-}\chi\text{-}\mathrm{pre}}$, or briefly as $\mathrm{EAR}_{\eta SE}^{\chi\text{-}\mathrm{pre}}$.

Following the convention introduced in footnote 1 in the present section, we let $\mathrm{AR}_{\eta SE}^{\chi\text{-}\mathrm{pre}}$ denote either $\mathrm{SAR}_{\eta SE}^{\chi\text{-}\mathrm{pre}}$ or $\mathrm{EAR}_{\eta SE}^{\chi\text{-}\mathrm{pre}}$. We will also occasionally briefly write $\mathrm{AR}_{\eta SE}^{\mathrm{pre}}$, where a suitable function

$$\chi\colon \mathrm{F}_{E}^{\mathrm{pre}} \longrightarrow \mathrm{F}_{S}^{\mathrm{total}}$$

will be understood.

We define the arithmetic $\mathrm{AR}_{\eta SE}^{\chi\text{-}\mathrm{pre}}$ as follows:

(I) The <u>functions</u> of $\mathrm{AR}_{\eta SE}^{\chi\text{-}\mathrm{pre}}$ are the functions in $\mathrm{F}_{\eta SE}^{\mathrm{scar}\text{-}\chi\text{-}\mathrm{pre}}$ (Definition 4).

(II) The <u>terms</u> of $\mathrm{AR}_{\eta SE}^{\chi\text{-}\mathrm{pre}}$ are those terms obtained by replacing any or all functions in terms of $\mathrm{AR}_{S}^{\mathrm{mon}}$ by functions in $F_{\eta SE}^{\text{scar-}\chi\text{-pre}}$.

(III) The <u>well-formed formulas</u> (or briefly <u>wffs</u>) of $AR_{\eta SE}^{\chi\text{-pre}}$ are the wffs in the set $\text{Wff}(AR_{\eta SE}^{\text{scar-}\chi\text{-pre}})$ defined in Proposition 13. We denote the set of wffs of $AR_{\eta SE}^{\chi\text{-pre}}$ briefly as

$$\text{Wff}(AR_{\eta SE}^{\chi\text{-pre}}).$$

(IV) The <u>axioms</u> of $AR_{\eta SE}^{\chi\text{-pre}}$ are those wffs ϖ of $AR_{\eta SE}^{\chi\text{-pre}}$ such that $\bar{\eta}(\varpi)$ is an axiom of AR_{E}^{pre}.

(V) The <u>axiom schemas</u> of $AR_{\eta SE}^{\chi\text{-pre}}$ are those of AR_{S}^{mon}.

(VI) The <u>inference rules</u> of $AR_{\eta SE}^{\chi\text{-pre}}$ are the inference rules of AR_{E}^{pre} (applied to wffs of $AR_{\eta SE}^{\chi\text{-pre}}$), where the successor set A_S is replaced by the semicarrier successor set A_{SE} (see Definition 4 above).

(VII) A wff $\varpi \in \text{Wff}(AR_{\eta SE}^{\chi\text{-pre}})$ that is obtained from the axioms of $AR_{\eta SE}^{\chi\text{-pre}}$ by a finite number of applications of the inference rules of $AR_{\eta SE}^{\chi\text{-pre}}$ is called a <u>theorem</u> of $AR_{\eta SE}^{\chi\text{-pre}}$.

We denote the set of theorems of $AR_{\eta SE}^{\chi\text{-pre}}$ as

$$\text{Thm}(AR_{\eta SE}^{\chi\text{-pre}}).$$

DEFINITION 9. Let S, ηSE, and $\bar{\eta}$ be as given in the paragraph preceding Definition 3, and let a function $\chi: F_{E}^{\text{rec}} \longrightarrow F_{S}^{\text{total}}$ be as given in Proposition 7.

In this definition we simultaneously define the <u>sentential semicarrier</u>-χ-<u>recursive arithmetic</u>, denoted as $SAR_{\eta SE}^{\text{scar-}\chi\text{-rec}}$, or briefly as $SAR^{\chi\text{-rec}}_{\eta SE}$, and the <u>equational semicarrier-χ-recursive arithmetic</u>, denoted as $EAR^{scar\text{-}\chi\text{-rec}}_{\eta SE}$, or briefly as $EAR^{\chi\text{-rec}}_{\eta SE}$.

Following the convention introduced in footnote 1 in the present section, we let $AR^{\chi\text{-rec}}_{\eta SE}$ denote either $SAR^{\chi\text{-rec}}_{\eta SE}$ or $EAR^{\chi\text{-rec}}_{\eta SE}$. We will also occasionally briefly write $AR^{rec}_{\eta SE}$, where a suitable function $\chi\colon F^{rec}_{E} \longrightarrow F^{total}_{S}$ will be understood.

We define the arithmetic $AR^{\chi\text{-rec}}_{\eta SE}$ as follows:

(I) The <u>functions</u> of $AR^{\chi\text{-rec}}_{\eta SE}$ are the functions in $F^{scar\text{-}\chi\text{-rec}}_{\eta SE}$ (Definition 5).

(II) The <u>terms</u> of $AR^{\chi\text{-rec}}_{\eta SE}$ are those terms obtained by replacing any or all functions in terms of AR^{cur}_{S} by functions in $F^{scar\text{-}\chi\text{-rec}}_{\eta SE}$.

(III) The <u>well-formed formulas</u> (or briefly <u>wffs</u>) of $AR^{\chi\text{-rec}}_{\eta SE}$ are the wffs in the set $\mathbf{Wff}(AR^{scar\text{-}\chi\text{-rec}}_{\eta SE})$ defined in Proposition 14. We denote the set of wffs of $AR^{\chi\text{-rec}}_{\eta SE}$ briefly as $\mathbf{Wff}(AR^{\chi\text{-rec}}_{\eta SE})$.

(IV) The <u>axioms</u> of $AR^{\chi\text{-rec}}_{\eta SE}$ are those wffs $\mathfrak{w}$ of $AR^{\chi\text{-rec}}_{\eta SE}$ such that $\overline{\eta}(\mathfrak{w})$ is an axiom of AR^{rec}_{E}.

(V) The <u>axiom schemas</u> of $SAR^{\chi\text{-rec}}_{\eta SE}$ (i.e., in the sentential case) are the axiom schemas S1, S2, S3, S4, and S6 of the sentential logic in Bourbaki [1968, Chapter I, §3, no. 1, and §5, no. 1] and the <u>axiom schemas</u> of $EAR^{\chi\text{-rec}}_{\eta SE}$ (i.e., in the equational case) are the two following (see §4.4 (footnote 1) for the notations $\dot{-}$ and x):

AXIOM SCHEMA I. For any $n \geq 1$ and $i \geq 0$, if γ_E is the regular primitive recursive function constant with Gödel number $\rho(n,i)$, then the following equation is an axiom of $\mathrm{EAR}_{\eta SE}^{\chi\text{-rec}}$ [5]:

$$\chi(\gamma_E)(x_1,\ldots,x_n,\chi(f_i^n)(x_1,\ldots,x_n)) = \#_S.$$

AXIOM SCHEMA II. For any $n \geq 1$ and $i \geq 0$, if γ_E is the regular primitive recursive function constant with Gödel number $\rho(n,i)$, then the following equation is an axiom of $\mathrm{EAR}_{\eta SE}^{\chi\text{-rec}}$:

$$(\chi(\sigma_E)(\#_S)\ \chi(\dot{-})\ (\chi(\sigma_E)(z)\ \chi(\dot{-})\ \chi(f_i^n)(x_1,\ldots,x_n)))\ \chi(\times)$$
$$(\chi(\sigma_E)(\#_S)\ \chi(\dot{-})\ \chi(\gamma_E)(x_1,\ldots,x_n,z)) = \#_S.$$

(VI) The inference rules of $\mathrm{AR}_{\eta SE}^{\chi\text{-rec}}$ are the inference rules of $\mathrm{AR}_E^{\mathrm{rec}}$ (applied to wffs of $\mathrm{AR}_{\eta SE}^{\chi\text{-rec}}$), where the successor set A_S is replaced by the semicarrier successor set A_{SE} (see Definition 4 above) [6].

---

(5) Axiom Schemas I and II are constructed so that $\overline{\eta}$ will map instances of these two axiom schemas to instances of Axiom Schemas I and II of $\mathrm{EAR}_E^{\mathrm{rec}}$ in §4.4 (cf. proof of Lemma 4 for Proposition 2 of §6.7).

(6) We note that although A_{SE} is (for convenience) defined in a definition that uses of the set $\mathbf{F}_E^{\mathrm{pre}}$, we may, without loss of generality, use the set A_{SE} in definitions that make use of the sets $\mathbf{F}_E^{\mathrm{rec}}$ and $\mathbf{F}_E^{\mathrm{parec}}$ (in which case we will assume that $A_{SE} \subseteq \mathbf{F}_{\eta SE}^{\mathrm{scar}\text{-}\chi\text{-rec}}$ or $A_{SE} \subseteq \mathbf{F}_{\eta SE}^{\mathrm{scar}\text{-}\chi\text{-parec}}$, as appropriate).

(VII) A wff $\mathfrak{w} \in \mathrm{Wff}(\mathrm{AR}_{\eta SE}^{\chi\text{-rec}})$ that is obtained from the axioms of $\mathrm{AR}_{\eta SE}^{\chi\text{-rec}}$ by a finite number of applications of the inference rules of $\mathrm{AR}_{\eta SE}^{\chi\text{-rec}}$ is called a <u>theorem</u> of $\mathrm{AR}_{\eta SE}^{\chi\text{-rec}}$.

We denote the set of theorems of $\mathrm{AR}_{\eta SE}^{\chi\text{-rec}}$ as

$$\mathrm{Thm}(\mathrm{AR}_{\eta SE}^{\chi\text{-rec}}).$$

DEFINITION 10. Let S, ηSE, and $\bar{\eta}$ be as given in the paragraph preceding Definition 3, and let a function

$$\chi\colon \mathrm{F}_{E}^{\mathrm{parec}} \longrightarrow \mathrm{F}_{S}^{\mathrm{partial}}$$

be as given in Proposition 8.

In this definition we simultaneously define the <u>sentential semicarrier</u>-χ-<u>partial</u>-<u>recursive</u> <u>arithmetic</u>, denoted as

$$\mathrm{SAR}_{\eta SE}^{\mathrm{scar}\text{-}\chi\text{-parec}},$$

or briefly as $\mathrm{SAR}_{\eta SE}^{\chi\text{-parec}}$, and the <u>equational</u> <u>semicarrier</u>-χ-<u>partial</u>-<u>recursive</u> <u>arithmetic</u>, denoted as $\mathrm{EAR}_{\eta SE}^{\mathrm{scar}\text{-}\chi\text{-parec}}$, or briefly as $\mathrm{EAR}_{\eta SE}^{\chi\text{-parec}}$.

Following the convention introduced in footnote 1 in the present section, we let $\mathrm{AR}_{\eta SE}^{\chi\text{-parec}}$ denote either $\mathrm{SAR}_{\eta SE}^{\chi\text{-parec}}$ or $\mathrm{EAR}_{\eta SE}^{\chi\text{-parec}}$. We will also occasionally briefly write $\mathrm{AR}_{\eta SE}^{\mathrm{parec}}$, where a suitable function $\chi\colon \mathrm{F}_{E}^{\mathrm{parec}} \longrightarrow \mathrm{F}_{S}^{\mathrm{partial}}$ will be understood.

We define the arithmetic $\mathrm{AR}_{\eta SE}^{\chi\text{-parec}}$ as follows:

(I) The <u>functions</u> of $\mathrm{AR}_{\eta SE}^{\chi\text{-parec}}$ are the functions in $\mathrm{F}_{\eta SE}^{\mathrm{scar}\text{-}\chi\text{-parec}}$ (Definition 6).

(II) The terms of $AR^{\chi\text{-parec}}_{\eta SE}$ are the terms obtained by replacing any or all functions in terms of AR^{pcur}_{S} by functions in $F^{scar\text{-}\chi\text{-parec}}_{\eta SE}$.

(III) The well-formed formulas (or briefly wffs) of $AR^{\chi\text{-parec}}_{\eta SE}$ are the wffs of the set $Wff(AR^{scar\text{-}\chi\text{-parec}}_{\eta SE})$ defined in Proposition 15. We denote the set of wffs of $AR^{\chi\text{-parec}}_{\eta SE}$ briefly as

$$Wff(AR^{\chi\text{-parec}}_{\eta SE}).$$

(IV) The axioms of $AR^{\chi\text{-parec}}_{\eta SE}$ are those wffs ϖ of $AR^{\chi\text{-parec}}_{\eta SE}$ such that $\bar{\eta}(\varpi)$ is an axiom of AR^{parec}_{E}.

(V) The axiom schemas of $SAR^{\chi\text{-parec}}_{\eta SE}$ (i.e., in the sentential case) are the axiom schemas S1, S2, S3, S4, and S6 of the sentential logic in Bourbaki [1968, Chapter I, §3, no. 1, and §5, no. 1] and the axiom schemas of $EAR^{\chi\text{-rec}}_{\eta SE}$ (i.e., in the equational case) are the four following (see §4.4 (footnote 1) for the notations $\dot{-}$ and $\times$):

AXIOM SCHEMA I. For any $n \geq 1$ and $i \geq 0$, if γ_E is the regular primitive recursive function constant with Gödel number $\rho(n, i)$, then the following equation is an axiom of $EAR^{\chi\text{-parec}}_{\eta SE}$ (7):

---

(7) Axiom Schemas I-IV are constructed so that $\bar{\eta}$ will map instances of these four axiom schemas to instances of Axiom Schemas I-IV of EAR^{parec}_{E} in §4.4 (cf. proof of Lemma 4 for Proposition 2 of §6.7, in the case of $EAR^{\chi\text{-rec}}_{\eta SE}$ and EAR^{rec}_{E}).

$$\chi(\gamma_E)(x_1,\ldots,x_n,\chi(f_i^n)(x_1,\ldots,x_n)) = \#_S.$$

AXIOM SCHEMA II. For any $n \geq 1$ and $i \geq 0$, if γ_E is the regular primitive recursive function constant with Gödel number $\rho(n,i)$, then the following equation is an axiom of $\mathrm{EAR}_{\eta SE}^{\chi\text{-parec}}$:

$$(\chi(\sigma_E)(\#_S)\ \chi(\dot{-})\ (\chi(\sigma_E)(z)\ \chi(\dot{-})\ \chi(f_i^n)(x_1,\ldots,x_n)))\ \chi(\times)$$
$$(\chi(\sigma_E)(\#_S)\ \chi(\dot{-})\ \chi(\gamma_E)(x_1,\ldots,x_n,z)) = \#_S.$$

AXIOM SCHEMA III. For any $n \geq 1$ and $i \geq 0$ and any constant terms $a_1, \ldots, a_n \in E$, if γ_E is the strictly partially regular primitive recursive function constant with Gödel number $\rho'(n,i)$, if the relation

$$(\exists y)(\gamma_E(a_1,\ldots,a_n,y) = 0)$$

is a theorem of RA [8], and if $a_1', a_2', \ldots, a_n'$ are the members of S such that

$$\eta(a_i') = a_i \quad (i = 1, 2, \ldots, n)\ [9],$$

then the following equation is an axiom of $\mathrm{EAR}_{\eta SE}^{\chi\text{-parec}}$:

---

[8] See §4.4.

[9] Note, $\eta: S \longrightarrow E$ is surjective by Definition 1.

$$\chi(\gamma_E)(a'_1,\ldots,a'_n,\chi(g^n_i)(a'_1,\ldots,a'_n)) = \#_S.$$

AXIOM SCHEMA IV. For any $n \geq 1$ and $i \geq 0$ and any constant terms $a_1, \ldots, a_n \in E$, if γ_E is the strictly partially regular primitive recursive function constant with Gödel number $\rho'(n,i)$, if the relation

$$(\exists y)(\gamma_E(a_1,\ldots,a_n,y) = 0)$$

is a theorem of RA, and if $a'_1, a'_2, \ldots, a'_n$ are the members of S such that

$$\eta(a'_i) = a_i \quad (i = 1, 2, \ldots, n),$$

then the following equation is an axiom of $\mathrm{EAR}^{\chi\text{-parec}}_{\eta SE}$:

$$(\chi(\sigma_E)(\#_S)\ \chi(\dot{-})\ (\chi(\sigma_E)(z)\ \chi(\dot{-})\ \chi(g^n_i)(a'_1,\ldots,a'_n)))\ \chi(\times)$$
$$(\chi(\sigma_E)(\#_S)\ \chi(\dot{-})\ \chi(\gamma_E)(a'_1,\ldots,a'_n,z)) = \#_S.$$

(VI) The inference rules of $\mathrm{AR}^{\chi\text{-parec}}_{\eta SE}$ are the inference rules of $\mathrm{AR}^{\mathrm{parec}}_E$ (applied to wffs of $\mathrm{AR}^{\chi\text{-parec}}_{\eta SE}$), where the successor set A_S is replaced by the semicarrier successor set A_{SE} (see Definition 4 above).

(VII) A wff $\mathfrak{w} \in \mathrm{Wff}(\mathrm{AR}^{\chi\text{-parec}}_{\eta SE})$ that is obtained from the axioms of $\mathrm{AR}^{\chi\text{-parec}}_{\eta SE}$ by a finite number of applications of the inference rules of $\mathrm{AR}^{\chi\text{-parec}}_{\eta SE}$ is called a theorem of

$$\mathrm{AR}^{\chi\text{-parec}}_{\eta SE}.$$

We denote the set of theorems of $AR^{\chi\text{-parec}}_{\eta SE}$ as

$$Thm(AR^{\chi\text{-parec}}_{\eta SE}).$$

PROPOSITION 18. Let S and ηSE be as given in the paragraph preceding Definition 3, let *** ∈ {pre, rec, parec}, and let χ be a function of the type given in Definitions 4-6. Then the set of functions $F^{scar\text{-}\chi\text{-}***}_{\eta SE}$, the set of wffs $Wff(AR^{\chi\text{-}***}_{\eta SE})$, and the set of theorems $Thm(AR^{\chi\text{-}***}_{\eta SE})$ of arithmetic $AR^{\chi\text{-}***}_{\eta SE}$ are all denumerably infinite.

Proof. Follows from Definitions 8-10 and Propositions 6-8 and 13-15. (To see that $Thm(AR^{\chi\text{-}***}_{\eta SE})$ is not less than denumerably infinite consider the wffs $x_1 = x_1$, $x_2 = x_2$, $x_3 = x_3$, ...)

PROPOSITION 19. Let S, ηSE, and $\bar{\eta}$ be as given in the paragraph preceding Definition 3, and let a function $\chi: F^{rec}_E \longrightarrow F^{total}_S$ be as given in Proposition 7. Then for each $\mathfrak{u} \in Wff(AR^{rec}_E)$ there exists a unique $\mathfrak{w}_0 \in Wff(AR^{scar\text{-}\chi\text{-}rec}_{\eta SE})$ such that $\bar{\eta}(\mathfrak{w}_0) = \mathfrak{u}$.

Proof. Let $\mathfrak{u} \in Wff(AR^{rec}_E)$. Then, letting $B_{\mathfrak{u}}$ be defined as in Proposition 10, we see by that reference that $B_{\mathfrak{u}} \neq \emptyset$, so that (still by Proposition 10) there exists a $\mathfrak{w} \in Wff^{total}_S$ such that $\bar{\eta}(\mathfrak{w}) = \mathfrak{u}$. We first prove that at least one wff $\mathfrak{w}_0$ of the desired type exists. We have two cases to consider:

First, if all of the functions in $\mathfrak{w}$ belong to $F^{scar\text{-}\chi\text{-}rec}_{\eta SE}$, then by Proposition 14 we have $\mathfrak{w} \in Wff(AR^{scar\text{-}\chi\text{-}rec}_{\eta SE})$, and we see that $\mathfrak{w}$ is at least one wff in $\mathrm{Wff}(\mathrm{AR}_{\eta SE}^{\mathrm{scar}\text{-}\chi\text{-}\mathrm{rec}})$ that is mapped by $\bar{\eta}$ to $\mathfrak{u}$.

Alternatively, assume that not all of the functions in $\mathfrak{w}$ belong to $\mathrm{F}_{\eta SE}^{\mathrm{scar}\text{-}\chi\text{-}\mathrm{rec}}$. Let f_1, f_2, ..., f_n be those functions occurring in $\mathfrak{w}$ (from left to right, where the f_i are possibly not distinct), in which case for some t, $1 \leq t \leq n$, and some i_1, i_2, ..., i_t, with $1 \leq i_1 < i_2 < \ldots < i_t \leq n$, we can let f_{i_1}, f_{i_2}, ..., f_{i_t} be those functions in $\mathfrak{w}$ that do not belong to $\mathrm{F}_{\eta SE}^{\mathrm{scar}\text{-}\chi\text{-}\mathrm{rec}}$.

Then by the definition of $\bar{\eta}$ (see Definition 2) we see that $\bar{\eta}(\mathfrak{w})$ is obtained by (a) replacing each occurrence of each constant $\mathbf{c}$ in $\mathfrak{w}$ by $\eta(\mathbf{c})$, and (b) for $i = 1, 2, \ldots, n$, replacing each occurrence of f_i in $\mathfrak{w}$ by $\eta[f_i]$. We now construct the desired wff $\mathfrak{w}_0$ as follows:

From $\eta[f_{i_1}]$, $\eta[f_{i_2}]$, ..., $\eta[f_{i_t}]$ occurring in $\mathfrak{u}$ (and $\mathfrak{u} \in \mathrm{Wff}(\mathrm{AR}_E^{\mathrm{rec}})$) we clearly have

$$\eta[f_{i_1}],\ \eta[f_{i_2}],\ \ldots,\ \eta[f_{i_t}] \in \mathrm{F}_E^{\mathrm{rec}},$$

which with Definition 5 yields

$$(1) \qquad \chi(\eta[f_{i_1}]),\ \chi(\eta[f_{i_2}]),\ \ldots,\ \chi(\eta[f_{i_t}]) \in \mathrm{F}_{\eta SE}^{\mathrm{scar}\text{-}\chi\text{-}\mathrm{rec}}.$$

We then define $\mathfrak{w}_0$ to be the formula obtained by replacing each occurrence of f_{i_k} in $\mathfrak{w}$ $(1 \leq k \leq t)$ by $\chi(\eta[f_{i_k}])$.

By Definition 5 we see that

(2) $$\eta[\chi(\eta[f_{i_k}])] = \eta[f_{i_k}] \quad (1 \leq k \leq t).$$

If $\mathfrak{w}$ is the string of symbols $s_1 s_2 \ldots s_m$ and $\mathfrak{w}_0$ is the string of symbols $s'_1 s'_2 \ldots s'_m$ (it is obvious by the construction of $\mathfrak{w}_0$ that $\mathfrak{w}$ and $\mathfrak{w}_0$ are of the same length) and we apply η to symbols as in relation (5) in the proof of Proposition 1 in §6.7, then

(3) $$\overline{\eta}(\mathfrak{w}) = \eta(s_1)\eta(s_2)\ldots\eta(s_m),$$

(4) $$\overline{\eta}(\mathfrak{w}_0) = \eta(s'_1)\eta(s'_2)\ldots\eta(s'_m).$$

It is clear that for each ℓ, $1 \leq \ell \leq m$, if s_ℓ isn't one of f_{i_1}, f_{i_2}, ..., f_{i_t}, then $s_\ell = s'_\ell$, so that $\eta(s_\ell) = \eta(s'_\ell)$.

On the other hand, for some ℓ, $1 \leq \ell \leq m$, assume that $s_\ell = f_{i_k}$, for some k, $1 \leq k \leq t$. Then by the construction of $\mathfrak{w}_0$ we have

$$s'_\ell = \chi(\eta[f_{i_k}]),$$

and with (2) we have

$$\eta[s'_\ell] = \eta[\chi(\eta[f_{i_k}])] = \eta[f_{i_k}] = \eta[s_\ell].$$

Thus with (3) and (4) we see that $\overline{\eta}(\mathfrak{w}) = \overline{\eta}(\mathfrak{w}_0)$, which with $\overline{\eta}(\mathfrak{w}) = \mathfrak{u}$ and $\mathfrak{u} \in \mathrm{Wff}(\mathrm{AR}_E^{\mathrm{rec}})$ (by the hypothesis of the proposition) yields $\overline{\eta}(\mathfrak{w}_0) \in \mathrm{Wff}(\mathrm{AR}_E^{\mathrm{rec}})$.

From $\mathfrak{w} \in \mathrm{Wff}_S^{\mathrm{total}}$ and the construction of $\mathfrak{w}_0$ we also have $\mathfrak{w}_0 \in \mathrm{Wff}_S^{\mathrm{total}}$, and then by part (b) of Definition 7 we have

(5) $$\mathfrak{w}_0 \in \mathfrak{W}_{\eta SE}^{\mathrm{scar\text{-}rec}}.$$

With (1) and the construction of $\mathfrak{w}_0$, it follows that all of the functions in $\mathfrak{w}_0$ belong to $F_{\eta SE}^{\text{scar-}\chi\text{-rec}}$, and so with (5) it follows by Proposition 14 that $\mathfrak{w}_0 \in \mathbf{Wff}(\mathrm{AR}_{\eta SE}^{\text{scar-}\chi\text{-rec}})$.

We thus see that in any case there exists at least one $\mathfrak{w}_0 \in \mathbf{Wff}(\mathrm{AR}_{\eta SE}^{\text{scar-}\chi\text{-rec}})$ such that $\bar{\eta}(\mathfrak{w}_0) = \mathfrak{u}$. We now show that this wff $\mathfrak{w}_0$ is uniquely determined. Let $\mathfrak{w}_0$ and $\mathfrak{u}$ be the strings of symbols $s_1 s_2 \ldots s_n$ and $s'_1 s'_2 \ldots s'_m$, respectively, for some $n \geq 1$ and $m \geq 1$. Then by relations (5) and (6) in the proof of Proposition 1 of §6.7 we have

$$\bar{\eta}(\mathfrak{w}_0) = \bar{\eta}(s_1 s_2 \ldots s_n) = \eta(s_1)\eta(s_2)\ldots\eta(s_n),$$

which with $\bar{\eta}(\mathfrak{w}_0) = \mathfrak{u} = s'_1 s'_2 \ldots s'_m$ yields

$$\eta(s_1)\eta(s_2)\ldots\eta(s_n) = s'_1 s'_2 \ldots s'_m,$$

in which case we clearly have $n = m$.

Now let $1 \leq i \leq n$. We will show that in each of the three possible cases the symbol s_i is uniquely determined:

Case 1. If s'_i (= $\eta(s_i)$) is a constant, then it is the root $\#_E$ (= 0) of space E ([10]), and by relation (5) in the proof of Proposition 1 of §6.7 we see that we must have $s_i = \#_S$.

Case 2. If s'_i (= $\eta(s_i)$) is a function, then by relation (5) in the proof of Proposition 1 in §6.7 we see that s_i must

---

([10]) All other constants of space E are represented formally by more than one symbol, i.e., $\sigma_E \circ \sigma_E \circ \ldots \circ \sigma_E(\#_E)$ (where σ_E is repeated some finite number of times).

also be a function. By Definition 5 of the present section we must have $s_i = \chi(s'_i)$, and again s_i is uniquely determined.

Case 3. If s'_i $(= \eta(s_i))$ is neither a constant nor a function, then by relation (5) in the proof of Proposition 1 of §6.7 we see that we must have $s_i = s'_i$.

Thus in each case the symbol s_i is uniquely determined, and it follows that the wff $\mathfrak{w}_0$ is itself uniquely determined.

§4.16. In this section, for any atomic infinite-dimensional nonzeroed semiological space S, any reversal Peano semicarrier space θES, and any *** ∈ {pre,rec,parec}, we define the arithmetic $\mathrm{AR}_{\theta ES}^{***}$ and then, for any Peano semicarrier space ηSE, establish various relationships among the arithmetics $\mathrm{AR}_{S}^{\mathrm{cur}}$, $\mathrm{AR}_{\eta SE}^{\mathrm{rec}}$, $\mathrm{AR}_{\theta ES}^{\mathrm{rec}}$, and $\mathrm{AR}_{E}^{\mathrm{rec}}$, as well as the corresponding results in the case of precursive and partial recursive semicarrier and reversal semicarrier arithmetics.

We begin with the following definition:

DEFINITION 0. Let S be a semiological space with structure $\mathscr{S}$, let S' be a set, and let $\eta: S \longrightarrow S'$ be a total injective function (in which case $\eta: S \longrightarrow \eta\langle S\rangle$ is a total bijective function with $\eta\langle S\rangle \subseteq S'$) [1]. By Bourbaki [**1968**, Chapter IV, §1, no. 5, Criterion CST5], there exists a unique semiological structure $\mathscr{S}'$ on $\eta\langle S\rangle$ such that η is an isomorphism of S onto $\eta\langle S\rangle$ (in the terminology of Bourbaki [**1968**, p. 268], $\mathscr{S}'$ is obtained by transporting the structure $\mathscr{S}$ to $\eta\langle S\rangle$ by means of η).

We then say that the set $\eta\langle S\rangle$ endowed with the semiological structure $\mathscr{S}'$ is the semiological space <u>induced on</u> $\eta\langle S\rangle$ <u>by</u> η.

In turn, if the semiological space S is nonzeroed, then:

---

[1] As noted in footnote [2] to the Theorem of Asser in §4.13, the notation $G\langle X\rangle$ $(= \{G(x) \mid x \in X\})$ is introduced in Bourbaki [**1968**, p. 77, Definition 3].

(a) We call arithmetic $AR^{mon}_{\eta\langle S\rangle}$ (§§4.6-4.7 and 4.11) the monocursive arithmetic induced on $\eta\langle S\rangle$ by η (in the event that $\eta\langle S\rangle$ is the set E endowed with the usual Peano semiological structure we see that $AR^{mon}_{\eta\langle S\rangle}$ is simply AR^{pre}_{E} (§§4.2 and 4.4) and we call it the precursive arithmetic induced on E by η).

(b) We call arithmetic $AR^{cur}_{\eta\langle S\rangle}$ (§§4.9 and 4.11) [resp. $AR^{pcur}_{\eta\langle S\rangle}$ (§§4.9 and 4.11)] the cursive [resp. partial cursive] arithmetic induced on $\eta\langle S\rangle$ by η (in the event that $\eta\langle S\rangle$ is the set E endowed with the usual Peano semiological structure we see that $AR^{cur}_{\eta\langle S\rangle}$ [resp. $AR^{pcur}_{\eta\langle S\rangle}$] is simply AR^{rec}_{E} (§§4.2 and 4.4) [resp. AR^{parec}_{E} (§§4.2 and 4.4)] and we call it the recursive [resp. partial recursive] arithmetic induced on E by η).

In turn, we introduce the relations of "subarithmetic of" and "isomorphic to" between arithmetics.

DEFINITION 1. Let AR_1 and AR_2 be arithmetics of any of the types defined here in Chapter 4 (i.e., precursive, recursive, partial recursive, monocursive, cursive, and partial cursive, as well as semicarrier and reversal semicarrier [2] arithmetics), where AR_1 and AR_2 need not be of the same type. Then we say that AR_1 is a subarithmetic of AR_2 if each of the following six conditions holds:

---

[2] Reversal semicarrier arithmetics are defined below in Definition 3.

(i) Each <u>function</u> of AR_1 is a restriction of some function of AR_2 ([3]).

(ii) For each <u>term</u> **t** of AR_1, if $\mathbf{f}_1, \mathbf{f}_2, \ldots, \mathbf{f}_n$ ($n \geq 0$) are the functions occurring in **t**, there exists a term **t'** of AR_2 and functions $\mathbf{f}'_1, \mathbf{f}'_2, \ldots, \mathbf{f}'_n$ such that

$$\mathbf{f}_i \text{ is a restriction of } \mathbf{f}'_i \ (i = 1, 2, \ldots, n)$$

and $\mathbf{f}'_1, \mathbf{f}'_2, \ldots, \mathbf{f}'_n$ are the functions occurring in **t'**, and, moreover, **t** can be obtained by replacing each $\mathbf{f}'_i$ by $\mathbf{f}_i$ ($i = 1, 2, \ldots, n$) throughout **t'**.

(iii) For each <u>wff</u> $\mathfrak{w}$ of AR_1, if $\mathbf{f}_1, \mathbf{f}_2, \ldots, \mathbf{f}_n$ ($n \geq 0$) are the functions occurring in $\mathfrak{w}$, there exists a wff $\mathfrak{w}'$ of AR_2 and functions $\mathbf{f}'_1, \mathbf{f}'_2, \ldots, \mathbf{f}'_n$ such that

$$\mathbf{f}_i \text{ is a restriction of } \mathbf{f}'_i \ (i = 1, 2, \ldots, n)$$

and $\mathbf{f}'_1, \mathbf{f}'_2, \ldots, \mathbf{f}'_n$ are the functions occurring in $\mathfrak{w}'$, and, moreover, $\mathfrak{w}$ can be obtained by replacing each $\mathbf{f}'_i$ by $\mathbf{f}_i$ ($i = 1, 2, \ldots, n$) in $\mathfrak{w}'$. We call $\mathfrak{w}$ a <u>restriction</u> of $\mathfrak{w}'$.

(iv) Each <u>axiom</u> of AR_1 is the restriction of some axiom of AR_2.

(v) all <u>axiom schemas</u> of AR_1 are axiom schemas of AR_2;

([3]) This includes the possibility that a function of AR_1 <u>be</u> a function of AR_2 (since a function is a restriction of itself (see Bourbaki [1968, Chapter II, §3, no. 5])).

(vi) If $\mathfrak{w}_1, \mathfrak{w}_2, \ldots, \mathfrak{w}_r$ $(r \geq 1)$ is a proof in AR_1, then there exists a proof $\mathfrak{w}'_1, \mathfrak{w}'_2, \ldots, \mathfrak{w}'_r$ in AR_2 such that

$\mathfrak{w}_i$ is a restriction of $\mathfrak{w}'_i$ $(i = 1, 2, \ldots, r)$.

(From the preceding it follows that every theorem of AR_1 is also a theorem of AR_2.)

PROPOSITION 1. For any arithmetics AR_1, AR_2, and AR_3:

(i) AR_1 is a subarithmetic of AR_1;

(ii) if AR_1 is a subarithmetic of AR_2 and AR_2 is a subarithmetic of AR_3, then AR_1 is a subarithmetic of AR_3.

If, in addition, each of the functions of AR_1 has the same domain and each of the functions of AR_2 has the same domain, then the following condition also holds:

(iii) if AR_1 is a subarithmetic of AR_2 and AR_2 is a subarithmetic of AR_1, then AR_1 and AR_2 are the same arithmetic ([4]).

(That is, in the case where each of the functions of AR_1 has the same domain and each of the functions of AR_2 has the same domain, the relation "is a subarithmetic of" is a reflexive, antisymmetric, and transitive relation, i.e., is an "inclusion" relation.)

([4]) That is, AR_1 and AR_2 have the same terms, functions, wffs, axioms, and proofs (and, hence, the same theorems).

Proof. Parts (i) and (ii) of the proposition follow immediately from Definition 1.

Assume, in addition, that

(*) all of the functions of AR_1 have the same domain and all of the functions of AR_2 have the same domain.

In turn, assume, that

(**) each of AR_1, AR_2 is a subarithmetic of the other.

Let f be a function of AR_1. Then by (**) and condition (i) of Definition 1 there exists a function f' of AR_2 such that $f \subseteq f'$ and, in turn, there exists a function f'' of AR_1 such that $f' \subseteq f''$. Thus $f \subseteq f' \subseteq f''$, so that $f \subseteq f''$. From this, with (*) (and with f and f'' both being functions of AR_1) one sees that $f = f''$, which with $f \subseteq f' \subseteq f''$ yields $f \subseteq f' \subseteq f$, so that $f = f'$. So with f' being a function of AR_2 it follows that f is a function of AR_2. Thus we see that all functions of AR_1 are functions of AR_2.

One easily modifies the preceding argument to a proof that all functions of AR_2 are functions of AR_1. Thus arithmetics AR_1 and AR_2 have the same functions.

Using similar arguments, one easily shows that the arithmetics AR_1 and AR_2 have the same terms (resp. wffs, resp. axioms, resp. proofs), and we see that arithmetics AR_1 and AR_2 are the same.

DEFINITION 1α. Let S be an atomic nonzeroed semiological space, let ηSE be a Peano semicarrier space, and let θES be a reversal Peano semicarrier space. Then (letting AR denote either SAR or EAR):

(i) For any $*** \in \{mon, cur, pcur\}$, we call S the underlying semiological space of the arithmetic AR_S^{***}.

(ii) For any $*** \in \{pre, rec, parec\}$, we call E the underlying semiological space of the arithmetic AR_E^{***}.

If, in addition, space S is infinite-dimensional, then:

(iii) For any $*** \in \{pre, rec, parec\}$, we call S the underlying semiological space of the arithmetic $AR_{\eta SE}^{***}$.

(iv) For any $*** \in \{pre, rec, parec\}$, we call the set $\theta\langle E\rangle$ endowed with the Peano structure transported to the set $\theta\langle E\rangle$ by the bijective mapping θ ([5]) the underlying semiological space of the arithmetic $AR_{\theta ES}^{***}$ (see Definition 3 below).

DEFINITION 2. As in Definition 1, let AR_1 and AR_2 be arithmetics of any of the types defined here in Chapter 4, and let S_1 and S_2 be the underlying semiological spaces of AR_1 and AR_2, respectively (see Definition 1α). Then we say that arithmetic AR_1 is isomorphic to arithmetic AR_2 if:

([5]) See Bourbaki [1968, Chapter IV, §1, no. 5, Criterion CST5], as well as Definition 0 of the present section.

(i) There exists an isomorphism of semiological space S_1 to semiological space S_2.

(ii) Arithmetics AR_1 and AR_2 are of the same type ([6]).

And in this case we refer to the isomorphism of S_1 to S_2 also as an <u>isomorphism of</u> AR_1 <u>to</u> AR_2.

We now extend Definition 2 of §4.15 to all semicarriers between the underlying semiological spaces of any arithmetics of the types defined here in Chapter 4 (cf. Definition 1):

DEFINITION 2α. Let AR_1 and AR_2 be arithmetics with underlying semiological spaces S_1 and S_2, respectively, with sets of functions F_1 and F_2, respectively, and sets of wffs $\mathbf{Wff}(AR_1)$ and $\mathbf{Wff}(AR_2)$, respectively, and let $\eta S_1 S_2$ be a semicarrier. Then we extend the mapping $\eta: S_1 \longrightarrow S_2$ to a partial function

$$\bar{\eta}: \mathbf{Wff}(AR_1) \longrightarrow \mathbf{Wff}(AR_2)$$

as follows: We define $\bar{\eta}(\mathfrak{w})$ for each $\mathfrak{w} \in \mathbf{Wff}(AR_1)$ such that for each function f occurring in $\mathfrak{w}$ we have $\eta[f] \in F_2$, and in that case we define $\bar{\eta}(\mathfrak{w})$ to be the formula obtained by (i) replacing each occurrence in $\mathfrak{w}$ of a constant c by $\eta(c)$, and

---

([6]) That is, both monocursive, both cursive, both partial cursive, or for some

$$*** \in \{\text{precursive}, \text{recursive}, \text{partial recursive}\},$$

both ***, both semicarrier ***, or both reversal semicarrier *** arithmetics.

(ii) replacing each occurrence in $\mathfrak{m}$ of a function f by $\eta[f]$ (which is, by hypothesis, in F_2).

We call $\bar{\eta}$ the *canonical extension of* η *to wffs*.

We note that if η is an isomorphism of S_1 to S_2 (and, *a fortiori*, if η is an isomorphism of arithmetic AR_1 to arithmetic AR_2), then $\bar{\eta}$: Wff(AR_1) $\longrightarrow$ Wff(AR_2) is a *total* function.

PROPOSITION 2. *For any arithmetics* AR_1, AR_2, *and* AR_3:

(i) AR_1 *is isomorphic to* AR_1.

(ii) *If* AR_1 *is isomorphic to* AR_2, *then* AR_2 *is isomorphic to* AR_1.

(iii) *If* AR_1 *is isomorphic to* AR_2 *and* AR_2 *is isomorphic to* AR_3, *then* AR_1 *is isomorphic to* AR_3.

(*That is, isomorphism between arithmetics is an equivalence relation. Therefore, instead of saying* "AR_1 *is isomorphic to* AR_2" *we can say* "AR_1 *and* AR_2 *are isomorphic*".)

Proof. Follows immediately from Definition 2.

PROPOSITION 2α. *For any arithmetics* AR_1 *and* AR_2 *of the types considered in* Definition 2, *with* S_1 *and* S_2 *being the underlying semiological space of* AR_1 *and* AR_2, *respectively, if* AR_1 *is isomorphic to* AR_2, *then there exists an isomorphism* γ *of* S_1 *to* S_2 *such that:*

(i) <u>if</u> AR_1 <u>and</u> AR_2 <u>are both precursive, both recursive, both partial recursive, both monocursive, both cursive, or both partial cursive, then</u> AR_2 <u>is the arithmetic induced on</u> S_2 <u>by</u> γ;

(ii) $\gamma[A_{S_1}] = A_{S_2}$;

(iii) <u>for all</u> $x \in S_1$, <u>we have</u> $[x \in R_{S_1} \leftrightarrow \gamma(x) \in R_{S_2}]$;

(iv) $\gamma\langle S_1\rangle = S_2$;

(v) <u>if</u> $\mathbf{F}_1$ <u>and</u> $\mathbf{F}_2$ <u>denote the sets of functions of arithmetics</u> AR_1 <u>and</u> AR_2, <u>respectively, then</u> $\gamma[\mathbf{F}_1] = \mathbf{F}_2$.

<u>In turn, recalling that</u> $\bar{\gamma}$ <u>denotes the canonical extension of</u> γ <u>to wffs</u> (Definition 2α), <u>we have the following three results</u>:

(vi) <u>if</u> $\mathbf{Wff}(AR_1)$ <u>and</u> $\mathbf{Wff}(AR_2)$ <u>denote the sets of wffs of arithmetics</u> AR_1 <u>and</u> AR_2, <u>respectively, then</u>

$$\bar{\gamma}\langle \mathbf{Wff}(AR_1)\rangle = \mathbf{Wff}(AR_2);$$

(vii) <u>if</u> $\mathrm{Axm}(AR_1)$ <u>and</u> $\mathrm{Axm}(AR_2)$ <u>denote the sets of axioms of arithmetics</u> AR_1 <u>and</u> AR_2, <u>respectively, then</u>

$$\bar{\gamma}\langle \mathrm{Axm}(AR_1)\rangle = \mathrm{Axm}(AR_2);$$

(viii) <u>if</u> $\mathrm{Thm}(AR_1)$ <u>and</u> $\mathrm{Thm}(AR_2)$ <u>denote the sets of theorems of arithmetics</u> AR_1 <u>and</u> AR_2, <u>respectively, then</u>

$$\bar{\gamma}\langle \mathrm{Thm}(AR_1)\rangle = \mathrm{Thm}(AR_2).$$

Proof. First, by Definition 2 there exists an isomorphism γ of semiological space S_1 to semiological space S_2, in which case $\gamma: S_1 \longrightarrow S_2$ is a total bijective function, so that $\gamma\langle S_1\rangle = S_2$.

So if AR_1 and AR_2 are both precursive, both recursive, both partial recursive, both monocursive, both cursive, or both partial cursive, so that for some *** $\in$ {pre,rec,parec} or some *** $\in$ {mon,cur,pcur} we can write

$$AR_1 = AR^{***}_{S_1} \qquad \text{and} \qquad AR_2 = AR^{***}_{S_2} = AR^{***}_{\gamma\langle S_1\rangle},$$

then by Definition 0 we see that $AR^{***}_{\gamma\langle S_1\rangle}$ (= $AR^{***}_{S_2}$) is the *** arithmetic induced on $\gamma\langle S_1\rangle$ (= S_2) by γ. This establishes item (i) of the proposition.

In turn, items (ii), (iii), and (iv) of the proposition follow from γ being an isomorphism of S_1 to S_2 (see Pogorzelski and Ryan [**1982**, Chapter I, §2, no. 3, Proposition 7]).

Next, recalling that the initial functions of a semiological space are the successor, projection, and constant functions (Pogorzelski and Ryan [**1985**, p. II.2.1.6, Definition 1]), one proves in a straightforward manner that

(1) <u>for every initial function</u> **f** <u>of arithmetic</u> AR_1, <u>the function</u> γ[**f**] <u>is an initial function of arithmetic</u> AR_2 <u>of the same type, and for every initial function</u> **f'** <u>of</u> AR_2, <u>there exists an initial function</u> **f** <u>of</u> AR_1 <u>of the same type such that</u> **f'** = γ[**f**].

In turn, with $\gamma: S_1 \longrightarrow S_2$ being a bijection, one proves in a straightforward manner that

(2) <u>for every function</u> **f** <u>of arithmetic</u> AR_1 <u>that is defined by the composition scheme, the function</u> $\gamma[\mathbf{f}]$ <u>of arithmetic</u> AR_2 <u>is defined by the composition scheme, and for every function</u> **f'** <u>of arithmetic</u> AR_2 <u>that is defined by the composition scheme, there exists a function</u> **f** <u>of arithmetic</u> AR_1 <u>that is defined by the composition scheme such that</u> $\gamma[\mathbf{f}] = \mathbf{f}'$.

Then, with $\gamma: S_1 \longrightarrow S_2$ being a bijection, one proves in a straightforward manner that

(3) <u>for every function</u> **f** <u>of arithmetic</u> AR_1 <u>that is defined by the monocursive scheme, the function</u> $\gamma[\mathbf{f}]$ <u>of arithmetic</u> AR_2 <u>is also defined by the monocursive scheme, and for every function</u> **f'** <u>of arithmetic</u> AR_2 <u>that is defined by the monocursive scheme, there exists a function</u> **f** <u>of arithmetic</u> AR_1 <u>that is defined by the monocursive scheme and such that</u> $\gamma[\mathbf{f}] = \mathbf{f}'$.

Next, using the definition of the canonical extension of a mapping to functions (Pogorzelski and Ryan [1982, Chapter 0, §1, no. 3]), along with the bijectivity of $\gamma: S_1 \longrightarrow S_2$, one proves in a straightforward manner that

(4) <u>for every function</u> **f** <u>of arithmetic</u> AR_1 <u>that is defined by the cursive scheme, the function</u> $\gamma[\mathbf{f}]$ <u>of arithmetic</u> AR_2 <u>is also defined by the cursive scheme, and for every function</u> **f'** <u>of arithmetic</u> AR_2 <u>that is defined by the cursive scheme, there exists a function</u> **f** <u>of arithmetic</u>

AR_1 <u>that is defined by the cursive scheme for which</u> $\gamma[f] = f'$.

In turn, referring to Pogorzelski and Ryan [**1985**, Chapter 0, §4, no. 4, Proposition 2] and using the definition of the canonical extension of a mapping to functions (Pogorzelski and Ryan [**1982**, Chapter 0, §1, no. 3]) and the bijectivity of $\gamma: S_1 \longrightarrow S_2$, one proves in a straightforward manner that (in the case where arithmetics AR_1 and AR_2 are both partial cursive)

(5) <u>for every function</u> **f** <u>of arithmetic</u> AR_1 <u>that is defined by the partial cursive scheme, the function</u> $\gamma[f]$ <u>of arithmetic</u> AR_2 <u>is also defined by the partial cursive scheme, and for every function</u> **f'** <u>of arithmetic</u> AR_2 <u>that is defined by the partial cursive scheme, there exists a function</u> **f** <u>of arithmetic</u> AR_1 <u>that is defined by the partial cursive scheme for which</u> $\gamma[f] = f'$.

With (1)-(5), item (v) of the proposition is established.

Items (vi)-(viii) of the proposition are proved using Definition 2α and the bijectivity of $\gamma: S_1 \longrightarrow S_2$ and

(I) observing that for any wff $\mathfrak{w} \in Wff(AR_1)$, one obtains $\bar{\gamma}(\mathfrak{w})$ by replacing each constant **c** occurring in $\mathfrak{w}$ by $\gamma(c)$ and replacing each function **f** occurring in $\mathfrak{w}$ by $\gamma[f]$, and

(II) observing that for any wff $\mathfrak{w}' \in Wff(AR_2)$, one obtains the wff $\mathfrak{w} \in Wff(AR_1)$ such that $\bar{\gamma}(\mathfrak{w}) = \mathfrak{w}'$ by replacing each constant **c'** occurring in $\mathfrak{w}'$ by the constant **c** of AR_1 for which

$\gamma(c) = c'$ and replacing each function f' occurring in $\mathfrak{w}'$ by the function f of AR_1 for which $\gamma[f] = f'$.

In particular, one establishes (vii) by obtaining

$$\bar{\gamma}\langle Axm(AR_1)\rangle \subseteq Axm(AR_2) \quad \text{and} \quad Axm(AR_2) \subseteq \bar{\gamma}\langle Axm(AR_1)\rangle$$

and one establishes (viii) by obtaining $\bar{\gamma}\langle Thm(AR_1)\rangle \subseteq Thm(AR_2)$ and $Thm(AR_2) \subseteq \bar{\gamma}\langle Thm(AR_1)\rangle$.

PROPOSITION 3. _Let S be a semiological space, let S' be a set, and let $\eta: S \longrightarrow S'$ be a total injective function. Then space S is isomorphic to the semiological space induced on $\eta\langle S\rangle$ by η._

Proof. By Definition 0 it follows that spaces S and $\eta\langle S\rangle$ are isomorphic in accordance with Bourbaki's definition of isomorphism of structures (Bourbaki [**1968**, Chapter IV, §1, no. 5]), and the proposition then follows by Pogorzelski and Ryan [**1982**, p. I.2.3.91, Proposition 32].

PROPOSITION 4. _Let S be an atomic nonzeroed semiological space and S' be a set, and let $\eta: S \longrightarrow S'$ be a total injective function. Then η is an isomorphism of the arithmetic_ AR_S^{mon} [resp. AR_S^{cur}, resp. AR_S^{pcur}] _to the monocursive_ [resp. _cursive_, resp. _partial cursive_] _arithmetic_ $AR_{\eta\langle S\rangle}^{mon}$ [resp. $AR_{\eta\langle S\rangle}^{cur}$, resp. $AR_{\eta\langle S\rangle}^{pcur}$] _induced on $\eta\langle S\rangle$ by η._

Proof. By Definition 0 we see that $\eta: S \longrightarrow \eta\langle S\rangle$ is a bijection. By Proposition 3 we see that η is an isomorphism of space S to the semiological space $\eta\langle S\rangle$ induced on the set $\eta\langle S\rangle$

by η. The proposition then follows by Definition 2.

DEFINITION 3. For any atomic infinite-dimensional nonzeroed semiological space S and any reversal Peano semicarrier space θES, we denote the precursive [resp. recursive, resp. partial recursive] arithmetic induced on $\theta\langle E\rangle$ by θ as $AR^{pre}_{\theta ES}$ [resp. $AR^{rec}_{\theta ES}$, resp. $AR^{parec}_{\theta ES}$] and we call it a <u>reversal</u> <u>semicarrier</u>-<u>precursive</u> [resp. <u>reversal</u> <u>semicarrier</u>-<u>recursive</u>, resp. <u>reversal</u> <u>semicarrier</u>-<u>partial</u>-<u>recursive</u>] <u>arithmetic</u>. (The terminology <u>reversal</u> <u>function</u> is introduced in Pogorzelski and Ryan [1982, p. I.2.7.1, Definition 2].)

Moreover, for $*** \in \{pre, rec, parec\}$,

(i) we let $F^{***}_{\theta ES}$ denote the set of <u>functions</u> of arithmetic $AR^{***}_{\theta ES}$;

(ii) we let $\mathbf{Wff}(AR^{***}_{\theta ES})$ denote the set of <u>wffs</u> of arithmetic $AR^{***}_{\theta ES}$;

(iii) we let $\mathrm{Thm}(AR^{***}_{\theta ES})$ denote the set of <u>theorems</u> of arithmetic $AR^{***}_{\theta ES}$.

PROPOSITION 5. <u>Let</u> S <u>be</u> <u>an</u> <u>atomic</u> <u>infinite</u>-<u>dimensional</u> <u>nonzeroed</u> <u>semiological</u> <u>space</u> <u>and</u> θES <u>be</u> <u>a</u> <u>reversal</u> <u>Peano</u> <u>carrier</u> <u>space</u>. <u>Then</u> <u>the</u> <u>reversal</u> <u>semicarrier</u> <u>map</u> θ <u>is</u> <u>an</u> <u>isomorphism</u> <u>of</u> <u>Peano</u> <u>space</u> E <u>to</u> <u>the</u> <u>semiological</u> <u>space</u> <u>induced</u> <u>on</u> $\theta\langle E\rangle$ <u>by</u> θ. <u>Moreover</u>, <u>for</u> $*** \in \{pre, rec, parec\}$, <u>the</u> <u>reversal</u> <u>semicarrier</u> <u>arithmetic</u> $AR^{***}_{\theta ES}$ <u>and</u> <u>the</u> <u>arithmetic</u> AR^{***}_{E} <u>are</u> <u>isomorphic</u>.

Proof. By Definition 1 of §4.15 it follows that $\theta: E \longrightarrow S$ is a total injective function. Then by Proposition 3 it follows that space E is isomorphic to the semiological space induced on $\theta\langle E\rangle$ by θ, with θ clearly being the isomorphism in question.

Then for $*** \in \{pre, rec, parec\}$, it follows by Proposition 4 that θ is an isomorphism of the arithmetic AR_E^{***} to the arithmetic $AR_{\theta\langle E\rangle}^{***}$ induced on $\theta\langle E\rangle$ by θ. By Definition 3 this is equivalent to saying that θ is an isomorphism of the arithmetic AR_E^{***} to the arithmetic $AR_{\theta ES}^{***}$. This completes the proof.

PROPOSITION 6. Let S be an atomic infinite-dimensional nonzeroed semiological space, let ηSE be a Peano semicarrier space, and let θES be a reversal Peano semicarrier space. Then we have the following relationships among certain of the various arithmetics that we have introduced here in Chapter 4:

First, in the monocursive/precursive case:

(ia) $AR_{\eta SE}^{pre}$ is a subarithmetic of AR_S^{mon};

(ib) $AR_{\theta ES}^{pre}$ is isomorphic to AR_E^{pre}.

In turn, in the cursive/recursive case:

(iia) $AR_{\eta SE}^{rec}$ is a subarithmetic of AR_S^{cur};

(iib) $AR_{\theta ES}^{rec}$ is isomorphic to AR_E^{rec}.

Lastly, in the partial cursive/partial recursive case:

(iiia) $AR^{parec}_{\eta SE}$ <u>is a subarithmetic of</u> AR^{pcur}_{S};

(iiib) $AR^{parec}_{\theta ES}$ <u>is isomorphic to</u> AR^{parec}_{E}.

<u>Proof</u>. Conditions (ia), (iia), and (iiia) are proved using Definitions 8-10 of §4.15 and Definition 1 of the present section. In turn, conditions (ib), (iib), and (iiib) are established in Proposition 5 of the present section.

EXAMPLE 1. Recall that the atomic infinite-dimensional commutative semiological space $\overline{F}$ (which is a number model of the commutative semiological space F) is defined in Pogorzelski and Ryan [1982, Chapter I, §4, no. 4, Example 4]. We define a function $\eta:\overline{F} \longrightarrow E$ by:

$$\eta(x) = x - 1 \quad (x \in \overline{F}),$$

and see that $\eta:\overline{F} \longrightarrow E$ is a bijection, so that $\eta^{-1}:E \longrightarrow \overline{F}$ is also a bijection and

$$\eta^{-1}\circ\eta(x) = x \quad (x \in \overline{F})$$

and

$$\overline{\eta}^{-1}\circ\overline{\eta}(\mathfrak{w}) = \mathfrak{w} \quad (\mathfrak{w} \in \mathbf{Wff}(AR^{mon}_{\overline{F}})).$$

Moreover, $\eta\overline{F}E$ is a Peano semicarrier space and $\eta^{-1}E\overline{F}$ is a reversal Peano semicarrier space.

We set $\chi = \eta^{-1}$ and leave it as an exercise for the reader to show that for $*** \in \{pre,rec,parec\}$, the semicarrier arithmetic $AR^{\chi-***}_{\eta\overline{F}E}$ and the reversal semicarrier arithmetic $AR^{***}_{\eta^{-1}E\overline{F}}$ are, in fact, the same arithmetic.

§4.17. In Definition 1 of the present section we define general recursive Peano semicarrier spaces and general recursive reversal Peano semicarrier spaces. We first present a discussion of Peano semicarrier spaces and reversal Peano semicarrier spaces showing why one must define general recursive Peano semicarrier spaces and general recursive reversal Peano semicarrier spaces only with respect to _number models_ of these semicarrier spaces.

For convenience, we state the following lemma:

LEMMA. _For any abstract Peano semicarrier space_ ηSE _and any abstract reversal Peano semicarrier space_ θES (_i.e._, _where_ S _is an abstract semiological space, although Peano space_ E _is assumed to be the set_ N _of natural numbers_), _if each element of_ S _is interpreted as a natural number, with each natural number being used and with different elements of_ S _being interpreted as different natural numbers, then one obtains a pair of number models_ $\eta^*\overline{S}E$ _and_ $\theta^*E\overline{S}$ _of_ ηSE _and_ θES [1], _where_ $\overline{S}$ = N _and where_ $\eta^*:\mathrm{N} \longrightarrow \mathrm{N}$ _is a total surjective function and_ $\theta^*:\mathrm{N} \longrightarrow \mathrm{N}$ _is a total injective function_.

Proof. In the terminology of Bourbaki [1968, Chapter IV, §1, no. 5], if ηSE and θES are considered as structures on the

---

(1) We point out that we will often speak of a pair of semicarrier spaces ηSE and θES as having number models $\eta^*\overline{S}E$ and $\theta^*E\overline{S}$, where it will be assumed that the elements of semiological space S have the same interpretations in both number models.

principal base set cl(S) with auxiliary base set E (= **N**), then $\eta^*\bar{S}E$ and $\theta^*E\bar{S}$ are the structures obtained by transporting the structures ηSE and θES to the set **N** by means of the bijection from S to **N** determined by the interpretations (mentioned in the statement of the lemma) of the members of S as natural numbers.

PROPOSITION 1. Let ηSE be an abstract Peano semicarrier space and let θES be an abstract reversal Peano semicarrier space (i.e., where S is an abstract semiological space, although Peano space E is assumed to be the set **N** of natural numbers). Then there exists a pair of number models $\eta^*\bar{S}E$ and $\theta^*E\bar{S}$ of ηSE and θES such that neither η^* nor θ^* is general recursive.

Proof. Let c_0, c_1, c_2, ... be the elements of S (denumerably infinite in number), so that

(1) $$S = \{c_0, c_1, c_2, \ldots\},$$

and let

(2) $$\eta = \{(c_0,n_0), (c_1,n_1), (c_2,n_2), \ldots\},$$

(3) $$\theta = \{(0,c_{i_0}), (1,c_{i_1}), (2,c_{i_2}), \ldots\},$$

where n_0, n_1, n_2, ... constitute the entire set **N** (although the n_i need not be distinct) and where the terms c_{i_0}, c_{i_1}, c_{i_2}, ... are distinct (although they need not constitute the entire set S), but where there are a denumerably infinite number of them (from θ being injective and having domain E).

By the Lemma we see that every one-to-one assignment of the natural numbers to the elements c_0, c_1, c_2, ... of S will yield a pair of number models $\eta^*\overline{S}E$ and $\theta^*E\overline{S}$ of ηSE and θES, where S = **N** (= E) and where η^* and θ^* are functions over the set **N** of natural numbers.

Using Cantor's diagonal method (and the fact that there are infinitely many of the terms c_0, c_1, c_2, ..., infinitely many of the terms c_{i_0}, c_{i_1}, c_{i_2}, ..., and infinitely many of the terms n_0, n_1, n_2, ...), we see that there are nondenumerably many pairs of number models $\eta^*\overline{S}E$ and $\theta^*E\overline{S}$ of the semicarrier spaces ηSE and θES.

Inasmuch as the systems of equations defining general recursive functions can be Gödel-numbered, there are only denumerably many of these systems of equations, and so there are only denumerably many general recursive functions.

Using these results one easily obtains the proposition.

Noting that the abstract Peano semicarrier space ϕFE and abstract reversal Peano semicarrier space ψEF are defined in §5.1 (first five pages), we have the following proposition:

PROPOSITION 2. *There exists a pair of number models* $\phi^*\overline{F}E$ *and* $\psi^*E\overline{F}$ *of* ϕFE *and* ψEF *such that the functions* ϕ^* *and* ψ^* *are both general recursive. On the other hand, there exists a pair of number models* $\phi^\circ\overline{\overline{F}}E$ *and* $\psi^\circ E\overline{\overline{F}}$ *of* ϕFE *and* ψEF *such that neither of the functions* ϕ°, ψ° *is general recursive*.

Proof. We have

$$F = \{\#_F,\ \sigma_1(\#_F),\ \sigma_2(\#_F),\ \sigma_1^2(\#_F),\ \sigma_3(\#_F),\ \sigma_1 \circ \sigma_2(\#_F),\ \ldots\}$$

and (cf. §5.1, Definitions 5 and 6 and paragraph preceding Definition 5) we see that $\phi = \lambda_F$ and that ϕ satisfies

$$\phi(\#_F) = 0$$

$$\phi(\sigma_{i_1} \circ \sigma_{i_2} \circ \ldots \circ \sigma_{i_n}(\#_F)) = n$$

$$(n \geq 1,\ \sigma_{i_1},\ \sigma_{i_2},\ \ldots,\ \sigma_{i_n} \in A_F)$$

and that ψ satisfies

$$\psi(0) = \#_F$$

$$\psi(n) = \sigma_n(\#_F) \quad (n \geq 1).$$

Then, interpreting the root $\#_F$ as the natural number 1 and (recalling that $p_1 = 2$, $p_2 = 3$, $p_3 = 5$, ... (Pogorzelski and Ryan [**1982**, Chapter 0, §0, no. 2]) interpreting each resolution

$$\sigma_{i_1} \circ \sigma_{i_2} \circ \ldots \circ \sigma_{i_n}(\#_F) \quad (n \geq 1)$$

in F as the natural number $p_{i_1} \cdot p_{i_2} \cdot \ldots \cdot p_{i_n}$, we obtain a number model $\phi^*\overline{F}E$ of the Peano semicarrier space ϕFE and a number model $\psi^*E\overline{F}$ of the reversal Peano semicarrier space ψEF such that the functions ϕ^* and ψ^* are both general recursive.

In turn, by Proposition 1 we see that there exists a pair of number models $\phi^\circ\overline{\overline{F}}E$ and $\psi^\circ E\overline{\overline{F}}$ of ϕFE and ψEF such that neither of the functions ϕ°, ψ° is general recursive. This completes the proof.

Noting that the abstract Peano semicarrier space αFE and abstract reversal Peano semicarrier space βEF are defined in §5.2 (first five pages), we have the following proposition:

PROPOSITION 3. <u>There exists a pair of number models</u> $\alpha^*\overline{F}E$ <u>and</u> $\beta^*E\overline{F}$ <u>of</u> αFE <u>and</u> βEF <u>such that the functions</u> α^* <u>and</u> β^* <u>are both general recursive. On the other hand, there exists a pair of number models</u> $\alpha^\circ\overline{\overline{F}}E$ <u>and</u> $\beta^\circ E\overline{\overline{F}}$ <u>of</u> αFE <u>and</u> βEF <u>such that neither of the functions</u> α°, β° <u>is general recursive</u>.

<u>Proof</u>. As in the proof of Proposition 2, we have

$$F = \{\#_F,\ \sigma_1(\#_F),\ \sigma_2(\#_F),\ \sigma_1^2(\#_F),\ \sigma_3(\#_F),\ \sigma_1\circ\sigma_2(\#_F),\ \ldots\}$$

and (§5.2, first five pages) we see that α satisfies

$$\alpha(\#_F) = 0$$

$$\alpha(\sigma_{i_1}\circ\sigma_{i_2}\circ\ \cdots\ \circ\sigma_{i_n}(\#_F)) = i_1 + i_2 + \ldots + i_n$$

$$(n \geq 1,\ \sigma_{i_1},\ \sigma_{i_2},\ \ldots,\ \sigma_{i_n} \in A_F)$$

and that β satisfies

$$\beta(0) = \#_F$$

$$\beta(n) = \sigma_n(\#_F) \qquad (n \geq 1).$$

Then, interpreting the root $\#_F$ as the natural number 1 and interpreting each resolution

$$\sigma_{i_1}\circ\sigma_{i_2}\circ\ \cdots\ \circ\sigma_{i_n}(\#_F) \qquad (n \geq 1)$$

in F as the natural number $p_{i_1} \cdot p_{i_2} \cdot \ldots \cdot p_{i_n}$, we obtain a number model $\alpha^* \overline{F}E$ of the Peano semicarrier space αFE and a number model $\beta^* E\overline{F}$ of the reversal Peano semicarrier space βEF such that the functions α^* and β^* are both general recursive.

In turn, by Proposition 1 we see that there exists a pair of number models $\alpha^\circ \overline{\overline{F}}E$ and $\beta^\circ E\overline{\overline{F}}$ of αFE and βEF such that neither of the functions α°, β° is general recursive. This completes the proof.

PROPOSITION 4. _Let ηSE be an abstract Peano semicarrier space and let θES be an abstract reversal Peano semicarrier space (i.e., where S is an abstract semiological space, although Peano space E is assumed to be the set_ **N** _of natural numbers). Then there exists a pair of number models $\eta^* \overline{S}E$ and $\theta^* E\overline{S}$ of ηSE and θES such that θ^* is general recursive. On the other hand, there exists a pair of number models $\eta^\circ \overline{\overline{S}}E$ and $\theta^\circ E\overline{\overline{S}}$ of ηSE and θES such that neither of the functions η°, θ° is general recursive._

Proof. Let ηSE be an abstract Peano semicarrier space, let θES be an abstract reversal Peano semicarrier space, and let

$$S = \{c_0, c_1, c_2, \ldots\}.$$

Then $\theta : E \longrightarrow S$ is injective (§4.15, Definition 1). So, using the ordered pair notation for functions, we can write the function θ in the form

$$\theta = \{(0, c_{i_0}), (1, c_{i_1}), (2, c_{i_2}), \ldots\},$$

where the terms c_{i_0}, c_{i_1}, c_{i_2}, ... are different from each other, in which case there is a denumerably infinite number of them.

One shows that there exists at least one pair of number models $\eta^*\overline{S}E$ and $\theta^*E\overline{S}$ of ηSE and θES such that θ^* is general recursive by considering the three cases:

(i) $\{c_{i_0}, c_{i_1}, c_{i_2}, \ldots\} = S$;

(ii) for some $n \geq 1$, there are exactly n members of S that do not occur among c_{i_0}, c_{i_1}, c_{i_2}, ...;

(iii) there is a denumerably infinite number of members of S not occurring among c_{i_0}, c_{i_1}, c_{i_2}, ...

In case (i) one interprets each c_{i_k} $(k \geq 0)$ as k. In case (ii) one interprets the members of S that do not occur among the c_{i_k} as 0, 1, ..., $n - 1$ and for each $k \geq 0$ interprets c_{i_k} as $k + n$. In case (iii) one interprets the members of S not occurring among the c_{i_k} as the odd natural numbers and then for each $k \geq 0$ interprets c_{i_k} as $2k$. One easily shows that in each of the cases (i) through (iii) one obtains a pair of number models $\eta^*\overline{S}E$ and $\theta^*E\overline{S}$ of ηSE and θES such that θ^* is a general recursive function (in the first case θ^* is the identity function on E, in the second case θ^* is the function $x + n$, and in the third case θ^* is the function $2x$).

On the other hand, we see by Proposition 1 that there

exists a pair of number models $\eta^{\circ}\bar{\bar{S}}E$ and $\theta^{\circ}E\bar{\bar{S}}$ of ηSE and θES such that neither of η°, θ° is general recursive. This completes the proof of Proposition 4.

DEFINITION 1. Let ηSE be an abstract Peano semicarrier space [resp. let θES be an abstract reversal Peano semicarrier space] (that is, where S is an abstract semiological space, although E is assumed to be the set **N** of natural numbers). Then we say that ηSE [resp. θES] is <u>general recursive with respect to number model</u> $\eta^{*}\bar{S}E$ [resp. <u>general recursive with respect to number model</u> $\theta^{*}E\bar{S}$] if η^{*} [resp. θ^{*}] is a general recursive function. We will often briefly say that an abstract Peano semicarrier space [resp. abstract reversal Peano semicarrier space] is <u>general recursive</u>, where it will be understood to be with respect to some number model.

DEFINITION 2. Noting that the abstract Peano semicarrier space αFE and the abstract reversal Peano semicarrier space βEF are defined in §5.2 (first five pages), we construct a pair of number models of αFE and βEF by considering E as the usual Peano number model and:

(i) interpreting the root $\#_F$ as the natural number 1, and

(ii) interpreting each resolution

$$\sigma_{i_1} \circ \sigma_{i_2} \circ \dots \circ \sigma_{i_n}(\#_F) \qquad (n \geq 1,\ \sigma_{i_1}, \sigma_{i_2}, \dots, \sigma_{i_n} \in A_F)$$

as the natural number $p_{i_1} \cdot p_{i_2} \cdot \ldots \cdot p_{i_n}$ (where $p_1 = 2$, $p_2 = 3$, $p_3 = 5$, ...).

We see that the number model of F thus obtained is the number model $\overline{F}$ defined in Pogorzelski and Ryan [**1982**, Chapter I, §4, no. 3, Example 4]. We can therefore write the number models of αFE and βEF thus obtained as $\alpha^*\overline{F}E$ and $\beta^*E\overline{F}$, where the functions α^* and β^* satisfy the recursions

$$\alpha^*(1) = 0$$

$$\alpha^*(p_{i_1} \cdot p_{i_2} \cdot \ldots \cdot p_{i_n}) = i_1 + i_2 + \ldots + i_n$$
$$(n \geq 1,\ i_1 \geq 1,\ i_2 \geq 1,\ \ldots,\ i_n \geq 1)$$

and

$$\beta^*(0) = 1$$

$$\beta^*(n) = p_n \quad (n \geq 1).$$

We easily see that α^* and β^* are general recursive, from which it follows (Definition 1) that the semicarrier spaces αFE and βEF are general recursive (with respect to the number models $\alpha^*\overline{F}E$ and $\beta^*E\overline{F}$).

§4.18. In this section we first define a family

$$(S_\mu)_{\mu \in \mathbf{N}}$$

which enumerates the abstract nonzeroed semiological spaces and their standard number models and then define a family

$$(\mathbf{S}_k)_{k \in \mathbf{N}}$$

that enumerates (i) the semiological spaces enumerated by the family $(S_\mu)_{\mu \in \mathbf{N}}$ and (ii) certain semicarriers $\overline{\eta S}_\mu E$ and $\overline{\theta} E \overline{S}_\mu$, where $\mu \geq 0$ and the semicarriers in question are number models, with $\overline{\eta}$ and $\overline{\theta}$ being either general recursive functions or restrictions of general recursive functions to S_μ.

Before giving $\mathcal{F}_{\mathbf{BS}}$-Definitions 1-3, we need the following notations and propositions:

We recall that $\mathbf{N} = \{0, 1, 2, \ldots\}$ and $\mathbf{N}^* = \mathbf{N} - \{0\}$. In turn, for each $r \geq 2$, for use in the following propositions and definitions, we define a set $\mathfrak{S}_r$ by:

$$\mathfrak{S}_r = \{p_1^{n_1} \cdot p_2^{n_2} \cdot \ldots \cdot p_r^{n_r} \mid n_1, n_2, \ldots, n_r \in \mathbf{N}\},$$

where $p_1, p_2, \ldots, p_r$ are the first r prime numbers in increasing order.

We recall (Pogorzelski and Ryan [1982, Chapter I, §4, no. 4, Examples 1-4, and no. 1, Propositions 11-14]) that

$$\overline{G}_k = \mathbf{N} \quad (k \geq 2), \qquad \overline{F}_k = \mathfrak{S}_k \quad (k \geq 1), \qquad \overline{G} = \overline{F} = \mathbf{N}^*.$$

In turn, we recall (§0.5 of the Introduction) that we are letting E denote both an abstract Peano space and the standard Peano number model. Lastly, we recall (Pogorzelski and Ryan [1982, Chapter I, §4, no. 4, Example 4]) that $\overline{F}$ denotes the multiplicative number model of the infinite-dimensional commutative semiological space F ([1]).

$\mathcal{F}_{BS}$-**DEFINITION 1.** For the remainder of this volume, let ηSE denote a general recursive Peano semicarrier space and let θES denote a general recursive reversal Peano semicarrier space, where space S is assumed to be atomic and nonzeroed, so that (cf. Definition 1 of §4.17) S is a Peano semiological space, either finite- or infinite-dimensional and either commutative or noncommutative.

$\mathcal{F}_{BS}$-**DEFINITION 2.** For the remainder of this volume we let $(S_\mu)_{\mu \in \mathrm{N}}$ be the family that enumerates all abstract nonzeroed semiological spaces and their standard number models (see Pogorzelski and Ryan [1982, Chapter I, §4, nos. 1 and 4]), according to the following conditions:

$$S_0 = F, \qquad S_1 = \overline{F}, \qquad S_2 = G, \qquad S_3 = \overline{G},$$

$S_4 = E$ (considered as the abstract Peano space),

([1]) This is in contrast with our denoting the partitional infinite-dimensional commutative semiological space (which is an additive number model of F) as F^+ (Pogorzelski and Ryan [1982, p. I.4.5.15]).

$S_5 = E$ (considered as the standard Peano number model),

$$S_{4n-2} = G_n \quad (n \geq 2), \qquad S_{4n-1} = \overline{G}_n \quad (n \geq 2),$$

$$S_{4n} = F_n \quad (n \geq 2), \qquad S_{4n+1} = \overline{F}_n \quad (n \geq 2).$$

Before defining the family $(S_k)_{k \in \mathbf{N}}$ we need the following eight propositions:

$\mathcal{F}_{BS}$-PROPOSITION 1. _For each_ $k \geq 2$, _the set of semicarriers_ $\overline{\eta G_k E}$ _such that_ $\overline{\eta}: \overline{G}_k \longrightarrow E$ _is a 1-place general recursive function is denumerable, and so we can let_ $\mathfrak{a}_1^{(k)}$, $\mathfrak{a}_2^{(k)}$, $\mathfrak{a}_3^{(k)}$, ... _be an enumeration of these semicarriers_ $\overline{\eta G_k E}$. _Therefore, there exists an enumeration_ $\mathfrak{A}_1$, $\mathfrak{A}_2$, $\mathfrak{A}_3$, ... _of all semicarriers_ $\overline{\eta G_k E}$, _where_ $k \geq 2$ _and where_ $\overline{\eta}$ _is a 1-place general recursive function_.

Proof. Kleene [**1952**, §57] proves that the systems of equations that define general recursive functions can be enumerated. Thus the set of general recursive functions is denumerable, and so with 1-place general recursive functions being functions $\eta: \mathbf{N} \longrightarrow \mathbf{N}$ and with $\overline{G}_k = \mathbf{N}$ $(k \geq 2)$ and $E = \mathbf{N}$ (see the paragraph preceding this proposition), we see that the set of 1-place general recursive functions $\overline{\eta}: \overline{G}_k \longrightarrow E$ $(k \geq 2)$ is denumerable. Then, since a subset of a countable set is also countable (Bourbaki [**1968**, Chapter III, §6, no. 4, Proposition 1]), noting Pogorzelski and Ryan [**1982**, Chapter I, §2, no. 4, Definition 1, and no. 3, Definition 1], we see that the set of semicarriers $\overline{\eta G_k E}$ $(k \geq 2)$ such that $\overline{\eta}$ is a 1-place general recursive function is countable. On the other hand, since there are obviously infinitely many general recursive bijections $\overline{\eta}:\overline{G}_k \longrightarrow E$ (2), it follows by Pogorzelski and Ryan [1982, Chapter I, §2, no. 4, Proposition 1] that there are infinitely many semicarriers $\overline{\eta G}_k E$ ($k \geq 2$) such that $\overline{\eta}$ is a 1-place general recursive function. It follows that for any $k \geq 2$, such semicarriers $\overline{\eta G}_k E$ can be enumerated, say by $\mathfrak{a}_1^{(k)}$, $\mathfrak{a}_2^{(k)}$, $\mathfrak{a}_3^{(k)}$, ...

Thus, letting $\mathfrak{A}_1 = \mathfrak{a}_1^{(2)}$, $\mathfrak{A}_2 = \mathfrak{a}_1^{(3)}$, $\mathfrak{A}_3 = \mathfrak{a}_2^{(2)}$, $\mathfrak{A}_4 = \mathfrak{a}_1^{(4)}$, $\mathfrak{A}_5 = \mathfrak{a}_2^{(3)}$, $\mathfrak{A}_6 = \mathfrak{a}_3^{(2)}$, $\mathfrak{A}_7 = \mathfrak{a}_1^{(5)}$, ..., we see that we can let $\mathfrak{A}_1$, $\mathfrak{A}_2$, $\mathfrak{A}_3$, ... be an enumeration of those semicarriers $\overline{\eta G}_k E$ such that $k \geq 2$ and $\overline{\eta}$ is a 1-place general recursive function (this is along the lines of Cantor's well-known enumeration of the rational numbers).

$\mathcal{F}_{BS}$**-PROPOSITION 2.** <u>For each</u> $k \geq 2$, <u>the set of semicarriers</u> $\overline{\theta E G}_k$ <u>such that</u> $\overline{\theta}:\mathrm{E} \longrightarrow \overline{G}_k$ <u>is a 1-place general recursive function is denumerable, and so we can let</u> $\mathfrak{b}_1^{(k)}$, $\mathfrak{b}_2^{(k)}$, $\mathfrak{b}_3^{(k)}$,

---

(2) Recalling that $\Delta_\mathbf{N}$ is the identity function on $\mathbf{N}$, set $\overline{\eta}_0 = \Delta_\mathbf{N}$ and, for $a = 1, 2, 3, \ldots$, let $\overline{\eta}_a$ be defined by:

$$\overline{\eta}_a(x) = x \quad (x \neq a \text{ and } x \neq a+1)$$

$$\overline{\eta}_a(a) = a + 1$$

$$\overline{\eta}_a(a+1) = a.$$

For $a = 1, 2, 3, \ldots$, the term $\overline{\eta}_a$ is the identity function on $\mathbf{N}$ with two values interchanged and is thus clearly a general recursive bijection $\overline{\eta}_a:\mathbf{N} \longrightarrow \mathbf{N}$.

... be an enumeration of these semicarriers $\overline{\theta E\overline{G}_k}$. Therefore, there exists an enumeration $\mathfrak{B}_1$, $\mathfrak{B}_2$, $\mathfrak{B}_3$, ... of all semicarriers $\overline{\theta E\overline{G}_k}$ such that $k \geq 2$ and such that $\overline{\theta}$ is a 1-place general recursive function.

Proof. The proof is obtained by interchanging the roles of $\overline{G}_k$ and E, replacing $\mathfrak{a}_i^{(k)}$ by $\mathfrak{b}_i^{(k)}$ ($k \geq 2$, $i \geq 1$), and replacing $\mathfrak{A}_i$ by $\mathfrak{B}_i$ ($i \geq 1$) in the proof of $\mathcal{F}_{\mathbf{BS}}$-Proposition 1.

$\mathcal{F}_{\mathbf{BS}}$**-PROPOSITION 3.** The set of semicarriers $\overline{\eta \overline{G} E}$ such that $\overline{\eta}:\overline{G} \longrightarrow E$ is the restriction to N^* of a general recursive function is denumerable, and so we can let $\mathfrak{C}_1$, $\mathfrak{C}_2$, $\mathfrak{C}_3$, ... be an enumeration of those semicarriers $\overline{\eta \overline{G} E}$ such that $\overline{\eta}$ is the restriction to N^* of a general recursive function.

Proof. As in the proof of $\mathcal{F}_{\mathbf{BS}}$-Proposition 1, we see that the set of general recursive functions is denumerable. Thus it is obvious that the set of 1-place functions $\overline{\eta}:\mathrm{N}^* \longrightarrow \mathrm{N}$ that are restrictions to N^* of general recursive functions is also denumerable. (In fact, for each such function $\overline{\eta}$, there exists a general recursive function $\overline{\eta}'$ such that, in set-theoretic notation, $\overline{\eta} = \overline{\eta}' - \{(0,\overline{\eta}'(0))\}$, so that there exists a bijection between the set of all such functions $\overline{\eta}$ and a subset of the set of general recursive functions.) From this, with $\overline{G} = \mathrm{N}^*$ and $E = \mathrm{N}$ (in the paragraph preceding $\mathcal{F}_{\mathbf{BS}}$-Proposition 1) it follows that the set of 1-place functions $\overline{\eta}:\overline{G} \longrightarrow E$ that are restrictions to N^* of general recursive functions is denumerable. Then since a subset of a countable set is also countable (Bourbaki [**1968**, Chapter III, §6, no. 4, Proposition 1]), noting Pogorzelski and Ryan [**1982**, Chapter I, §2, no. 4, Definition 1, and no. 3, Definition 1], we see that the set of semicarriers $\overline{\eta G E}$ such that $\overline{\eta}$ is the restriction to $\mathbf{N}^*$ of a 1-place general recursive function is countable. On the other hand, since there are obviously infinitely many 1-place general recursive bijections $\overline{\eta}:\overline{G} \longrightarrow E$, with $\overline{\eta}$ being the restriction to $\mathbf{N}^*$ of a general recursive function [3], it follows by Pogorzelski and Ryan [**1982**, Chapter I, §2, no. 4, Proposition 1] that there are infinitely many semicarriers $\overline{\eta G E}$ such that $\overline{\eta}$ is the restriction to $\mathbf{N}^*$ of a 1-place general recursive function. Thus such semicarriers $\overline{\eta G E}$ can be enumerated, say by $\mathfrak{C}_1$, $\mathfrak{C}_2$, $\mathfrak{C}_3$, ...

$\mathscr{F}_{\mathbf{BS}}$**-PROPOSITION 4.** <u>The set of semicarriers $\overline{\theta E \overline{G}}$ such that $\overline{\theta}: E \longrightarrow \overline{G}$ is a 1-place general recursive function is denumerable, and so we can let $\mathfrak{D}_1$, $\mathfrak{D}_2$, $\mathfrak{D}_3$, ... be an enumeration of those semicarriers $\overline{\theta E \overline{G}}$ such that $\overline{\theta}$ is a general recursive function</u>.

<u>Proof</u>. As in the proof of $\mathscr{F}_{\mathbf{BS}}$-Proposition 1, we see that the set of general recursive functions is denumerable. Thus it is

---

(3) Let $\overline{\eta}_0:\mathbf{N}^* \longrightarrow \mathbf{N}$ be the bijection that satisfies $\overline{\eta}_0(x) = x - 1$ $(x \in \mathbf{N}^*)$. Then for $a = 1, 2, 3, \ldots,$ define $\overline{\eta}_a:\mathbf{N}^* \longrightarrow \mathbf{N}$ by the three equations $\overline{\eta}_a(x) = \overline{\eta}_0(x)$ $(x \neq a,$ $x \neq a+1)$; $\overline{\eta}_a(a) = a$; $\overline{\eta}_a(a+1) = a - 1$.

obvious that the set of all 1-place general recursive functions $\overline{\theta}:\mathrm{N} \longrightarrow \mathrm{N}^*$ is also denumerable. From this, with $E = \mathrm{N}$ and $\overline{G} = \mathrm{N}^*$ (in the paragraph preceding $\mathscr{F}_{\mathrm{BS}}$-Proposition 1), it follows that the set of 1-place general recursive functions $\overline{\theta}:E \longrightarrow \overline{G}$ is denumerable. Then since a subset of a countable set is also countable (Bourbaki [1968, Chapter III, §6, no. 4, Proposition 1]), noting Pogorzelski and Ryan [1982, Chapter I, §2, no. 4, Definition 1, and no. 3, Definition 1], we see that the set of semicarriers $\overline{\theta}E\overline{G}$ such that $\overline{\theta}$ is general recursive is denumerable. On the other hand, since there are obviously infinitely many 1-place general recursive bijections $\overline{\theta}:E \longrightarrow \overline{G}$ (4), it follows by Pogorzelski and Ryan [1982, Chapter I, §2, no. 4, Proposition 1] that there are infinitely many semicarriers $\overline{\theta}E\overline{G}$, with $\overline{\theta}$ general recursive. Thus such semicarriers $\overline{\theta}E\overline{G}$ can be enumerated, say by $\mathfrak{D}_1$, $\mathfrak{D}_2$, $\mathfrak{D}_3$, ...

$\mathscr{F}_{\mathrm{BS}}$-PROPOSITION 5. _The set of semicarriers $\overline{\eta}\overline{F}E$ such that $\overline{\eta}:\overline{F} \longrightarrow E$ is the restriction to N^* of a general recursive function is denumerable, and so we can let $\mathfrak{C}_1$, $\mathfrak{C}_2$, $\mathfrak{C}_3$, ... be an enumeration of those semicarriers $\overline{\eta}\overline{F}E$ such that $\overline{\eta}$ is the restriction to N^* of a general recursive function._

---

(4) Define $\overline{\theta}_0:\mathrm{N} \longrightarrow \mathrm{N}^*$ by $\overline{\theta}_0(x) = x + 1$ $(x \in \mathrm{N})$, which is obviously general recursive. Then, for $a = 1, 2, 3, \ldots,$ define $\overline{\theta}_a:\mathrm{N} \longrightarrow \mathrm{N}^*$ by the three equations $\overline{\theta}_a(x) = \overline{\theta}_0(x)$ $(x \neq a,\ x \neq a+1)$; $\overline{\theta}_a(a) = \overline{\theta}_0(a+1)$; $\overline{\theta}_a(a+1) = \overline{\theta}_0(a)$.

Proof. One obtains the proof by replacing $\overline{G}$ by $\overline{F}$ and replacing $\mathfrak{C}_k$ by $\mathfrak{G}_k$ $(k \geq 1)$ in the proof of $\mathcal{F}_{\mathrm{BS}}$-Proposition 5.

$\mathcal{F}_{\mathrm{BS}}$-**PROPOSITION 6.** *The set of semicarriers $\overline{\theta}E\overline{F}$ such that $\overline{\theta}: E \longrightarrow \overline{F}$ is a 1-place general recursive function is denumerable, and so we can let $\mathfrak{Z}_1$, $\mathfrak{Z}_2$, $\mathfrak{Z}_3$, ... be an enumeration of those semicarriers $\overline{\theta}E\overline{F}$ such that $\overline{\theta}$ is a 1-place general recursive function.*

Proof. One obtains the proof by replacing $\overline{G}$ by $\overline{F}$ and replacing $\mathfrak{D}_k$ by $\mathfrak{Z}_k$ $(k \geq 1)$ in the proof of $\mathcal{F}_{\mathrm{BS}}$-Proposition 6.

$\mathcal{F}_{\mathrm{BS}}$-**PROPOSITION 7.** *The set of semicarriers $\overline{\eta F_k}E$ $(k \geq 1)$ such that $\overline{\eta}: \overline{F}_k \longrightarrow E$ is the restriction to $\mathfrak{S}_k$ of a 1-place general recursive function is denumerable, and so we can let $\mathfrak{g}_1^{(k)}$, $\mathfrak{g}_2^{(k)}$, $\mathfrak{g}_3^{(k)}$, ... be an enumeration of those semicarriers $\overline{\eta F_k}E$ such that $\overline{\eta}$ is the restriction to $\mathfrak{S}_k$ of a 1-place general recursive function. Therefore, there exists an enumeration $\mathfrak{G}_1$, $\mathfrak{G}_2$, $\mathfrak{G}_3$, ... of all semicarriers $\overline{\eta F_k}E$ such that $k \geq 1$ and $\overline{\eta}$ is the restriction to $\mathfrak{S}_k$ of a 1-place general recursive function.*

Proof. Let $k \geq 1$. Then one, for the most part, follows the proof of $\mathcal{F}_{\mathrm{BS}}$-Proposition 3, replacing N^* by $\mathfrak{S}_k$, replacing the equation $\overline{\eta} = \overline{\eta}' - \{(x, \overline{\eta}'(x))\}$ by the equation

$$\overline{\eta} = \overline{\eta}' - \{(x, \overline{\eta}'(x)) \mid x \in \mathrm{N} - \mathfrak{S}_k\},$$

enumerating the semicarriers $\overline{\eta F_k}E$ such that $\overline{\eta}$ is the restriction to $\mathfrak{S}_k$ of a 1-place general recursive function by

$\mathfrak{g}_1^{(k)}$, $\mathfrak{g}_2^{(k)}$, $\mathfrak{g}_3^{(k)}$, ..., enumerating the semicarriers $\overline{\eta F_k E}$ (for all $k \geq 1$) by $\mathfrak{G}_1$, $\mathfrak{G}_2$, $\mathfrak{G}_3$, ..., and using the following proof that there are infinitely many 1-place general recursive bijections $\overline{\eta}:F_k \longrightarrow E$, with $\overline{\eta}$ being the restriction to $\mathfrak{S}_k$ of a 1-place general recursive function:

Using Goodstein's [**1957**, p. 64] limited existential operator, we define general recursive relations $x \in \mathfrak{S}_k$ and $x \notin \mathfrak{S}_k$ by:

$$x \in \mathfrak{S}_k \quad =_{\text{Def}} \quad \neg\, \exists_m^x [k < m \quad \& \quad p_m | x],$$

$$x \notin \mathfrak{S}_k \quad =_{\text{Def}} \quad \exists_m^x [k < m \quad \& \quad p_m | x].$$

In turn, using Goodstein [**1957**, p. 84, Example 3.6], we define the limited maximal operator G_x^n so that for any general recursive predicate $P(x)$, for any $n \geq 0$ the expression $G_x^n P(x)$ yields the greatest x not exceeding n such that $P(x)$ holds, if such an x exists; if no such x exists, then $G_x^n P(x) = n$.

We define a function $\overline{\eta}_0:\mathbf{N} \longrightarrow \mathbf{N}$ as follows:

$$\overline{\eta}_0(0) = 0$$

$$\overline{\eta}_0(1) = 0$$

$$\overline{\eta}_0(a) = \begin{cases} 0 \quad (\text{if } a \notin \mathfrak{S}_k) \\ \\ \overline{\eta}_0(G_x^a[x < a \quad \& \quad x \in \mathfrak{S}_k]) + 1 \\ \qquad\qquad\qquad (\text{if } a \in \mathfrak{S}_k \quad \& \quad a > 1). \end{cases}$$

We see that $\overline{\eta}_0:\mathbf{N} \longrightarrow \mathbf{N}$ is a 1-place general recursive function. In turn, for each $a \geq 1$ we define a function

$\overline{\eta}_a: \mathrm{N} \longrightarrow \mathrm{N}$ by the equations:

$$\overline{\eta}_a(x) = \overline{\eta}_0(x) \quad (x \neq p_1^a \cdot p_2^a \cdot \ldots \cdot p_k^a,\ x \neq p_1^{a+1} \cdot p_2^{a+1} \cdot \ldots \cdot p_k^{a+1})$$

$$\overline{\eta}_a(p_1^a \cdot p_2^a \cdot \ldots \cdot p_k^a) = \overline{\eta}_0(p_1^{a+1} \cdot p_2^{a+1} \cdot \ldots \cdot p_k^{a+1})$$

$$\overline{\eta}_a(p_1^{a+1} \cdot p_2^{a+1} \cdot \ldots \cdot p_k^{a+1}) = \overline{\eta}_0(p_1^a \cdot p_2^a \cdot \ldots \cdot p_k^a).$$

We see that $\overline{\eta}_a: \mathrm{N} \longrightarrow \mathrm{N}$ is a 1-place general recursive function (in fact, $\overline{\eta}_a$ is simply the function $\overline{\eta}_0$ with two values interchanged). We also see that each of $\overline{\eta}_0$, $\overline{\eta}_1$, $\overline{\eta}_2$, ... is distinct, so that there are infinitely many of them.

Now for each $a \geq 0$ we let $\overline{\eta}'_a$ be the restriction of $\overline{\eta}_a$ to $\mathfrak{S}_k$ and we see that $\overline{\eta}'_a: \mathfrak{S}_k \longrightarrow \mathrm{N}$ is a bijection (in particular, $\overline{\eta}'_0$ is an isomorphism of $\mathfrak{S}_k$ onto N (Bourbaki [1968, Chapter III, §1, no. 3, last paragraph]), with both sets being well-ordered in the natural way).

$\mathscr{F}_{\mathbf{BS}}$-**PROPOSITION 8.** _The set of semicarriers_ $\overline{\theta}E\overline{F}_k$ $(k \geq 1)$ _such that_ $\overline{\theta}: E \longrightarrow \overline{F}_k$ _is a 1-place general recursive function is denumerable, and so we can let_ $\mathfrak{b}_1^{(k)}$, $\mathfrak{b}_2^{(k)}$, $\mathfrak{b}_3^{(k)}$, ... _be an enumeration of those semicarriers_ $\overline{\theta}E\overline{F}_k$ _such that_ $\overline{\theta}$ _is a 1-place general recursive function. Therefore, there exists an enumeration_ $\mathfrak{B}_1$, $\mathfrak{B}_2$, $\mathfrak{B}_3$, ... _of all semicarriers_ $\overline{\theta}E\overline{F}_k$ _where_ $k \geq 1$ _and_ $\overline{\theta}$ _is a 1-place general recursive function._

Proof. One obtains the proof by making the obvious changes to the proof of $\mathscr{F}_{\mathbf{BS}}$-Proposition 4.

$\mathcal{F}_{BS}$**-DEFINITION 3.** For the remainder of this volume we let $(\mathfrak{S}_k)_{k \in \mathbf{N}}$ be a family that enumerates (i) the semiological spaces enumerated by the family $(S_\mu)_{\mu \in \mathbf{N}}$ (see $\mathcal{F}_{BS}$-Definition 2) and (ii) certain semicarriers $\overline{\eta S}_\mu E$ and $\overline{\theta E S}_\mu$, where $\mu \geq 0$ and the semicarriers in question are number models (with $\overline{\eta}$ and $\overline{\theta}$ being either general recursive functions or restrictions of general recursive functions). For each $k \geq 0$, recalling the notations $\mathfrak{A}_k$, $\mathfrak{B}_k$, ..., $\mathfrak{H}_k$ introduced in $\mathcal{F}_{BS}$-Propositions 1-8, we define:

$$\mathfrak{S}_{9k} = S_k, \qquad \mathfrak{S}_{9k+1} = \mathfrak{A}_k, \qquad \mathfrak{S}_{9k+2} = \mathfrak{B}_k,$$

$$\mathfrak{S}_{9k+3} = \mathfrak{C}_k, \qquad \mathfrak{S}_{9k+4} = \mathfrak{D}_k, \qquad \mathfrak{S}_{9k+5} = \mathfrak{E}_k,$$

$$\mathfrak{S}_{9k+6} = \mathfrak{F}_k, \qquad \mathfrak{S}_{9k+7} = \mathfrak{G}_k, \qquad \mathfrak{S}_{9k+8} = \mathfrak{H}_k.$$

At times, for typographical convenience, for any $k \geq 0$ we will write $\mathfrak{S}$ instead of $\mathfrak{S}_k$.

$\mathcal{F}_{BS}$**-DEFINITION 4.** For the remainder of this volume:

(i) We let $\mathrm{AR}_E^{\mathrm{rec}}$ denote either sentential recursive arithmetic $\mathrm{SAR}_E^{\mathrm{rec}}$ or equational recursive arithmetic $\mathrm{EAR}_E^{\mathrm{rec}}$.

(ii) For any Peano semicarrier space ηSE and any reversal Peano semicarrier space θES, we let $\mathrm{AR}_{SE}^{\mathrm{rec}}$ denote either sentential recursive arithmetic $\mathrm{SAR}_{SE}^{\mathrm{rec}}$ or equational recursive arithmetic $\mathrm{EAR}_{SE}^{\mathrm{rec}}$, where SE denotes either ηSE or θES. (This is in contrast to the use of the notation SE in the notation

$\mathbf{G}_{SE}^{\chi\text{-}***}$ which we introduce in §5.3 and the use of the notation *FE* in the notation $\mathcal{V}_{FE}^{\chi\text{-rec}}$ which we introduce in §5.7 and the terminology "*FE*-range" and notation range_{FE} which we also introduce in §5.7, where the notations *SE* and *FE* are simply a part of the expression in question and do not denote the conjunction of a Peano semicarrier and a reversal Peano semicarrier.)

§4.19. In this section and in §4.20, for suitable Peano semicarrier ηSE and total function $\chi:\mathrm{F}_E^{rec} \longrightarrow \mathrm{F}_S^{total}$, we prove certain relationships between the set of wffs of AR_E^{rec} and the set of wffs of $\mathrm{AR}_{\eta SE}^{\chi\text{-}rec}$ and certain relationships between the set of theorems of AR_E^{rec} and the set of theorems of $\mathrm{AR}_{\eta SE}^{\chi\text{-}rec}$.

$\mathcal{F}_{BS}$-**PROPOSITION 1.** Let S be an atomic infinite-dimensional nonzeroed semiological space, let ηSE be a Peano semicarrier space, let $\bar{\eta}$ be the canonical recursive-extension of η to wffs, and let $\chi:\mathrm{F}_E^{rec} \longrightarrow \mathrm{F}_S^{total}$ be a total function of the type specified in Definition 5 of §4.15. Then

$$\bar{\eta}(\mathrm{Wff}(\mathrm{AR}_{\eta SE}^{\chi\text{-}rec})) = \mathrm{Wff}(\mathrm{AR}_E^{rec}).$$

(We at times briefly express this result as

$$\bar{\eta}(\mathrm{Wff}(\mathrm{AR}_{SE}^{rec})) = \mathrm{Wff}(\mathrm{AR}_E^{rec}).)$$

Proof. As in Proposition 10 of §4.15, let a family

$$\mathcal{W} = (B_{\mathfrak{u}})_{\mathfrak{u} \in \mathrm{Wff}(\mathrm{AR}_E^{rec})}$$

be defined, where

(1) for each $\mathfrak{u} \in \mathrm{Wff}(\mathrm{AR}_E^{rec})$ we have

$$B_{\mathfrak{u}} = \{\mathfrak{w} \mid \mathfrak{w} \in \mathrm{Wff}_S^{total} \text{ and } \bar{\eta}(\mathfrak{w}) = \mathfrak{u}\}.$$

First, let $\mathfrak{u} \in \bar{\eta}(\mathrm{Wff}(\mathrm{AR}_{\eta SE}^{\chi\text{-}rec}))$, in which case for some $\mathfrak{w} \in \mathrm{Wff}(\mathrm{AR}_{\eta SE}^{\chi\text{-}rec})$ we have $\mathfrak{u} = \bar{\eta}(\mathfrak{w})$.

By condition (III) in Definition 9 of §4.15 we have

$$\mathrm{Wff}(\mathrm{AR}_{\eta SE}^{\chi\text{-rec}}) = \mathrm{Wff}(\mathrm{AR}_{\eta SE}^{\text{scar-}\chi\text{-rec}}),$$

and so we have

$$\mathfrak{w} \in \mathrm{Wff}(\mathrm{AR}_{\eta SE}^{\text{scar-}\chi\text{-rec}}).$$

Then by Proposition 14 of §4.15,

(2) $\mathfrak{w} \in \mathfrak{W}_{\eta SE}^{\text{scar-rec}}$ <u>and all of the functions in</u> $\mathfrak{w}$ <u>belong to</u> $\mathrm{F}_{\eta SE}^{\text{scar-}\chi\text{-rec}}$.

By (2), $\mathfrak{w}$ is a semicarrier-recursive S-wff with respect to ηSE (by part (b) of Definition 7 of §4.15), which by the same reference implies $\overline{\eta}(\mathfrak{w}) \in \mathrm{Wff}(\mathrm{AR}_E^{\text{rec}})$. Thus with $\mathfrak{u} = \overline{\eta}(\mathfrak{w})$ we have $\mathfrak{u} \in \mathrm{Wff}(\mathrm{AR}_E^{\text{rec}})$, and so we have proved that

$$\overline{\eta}(\mathrm{Wff}(\mathrm{AR}_{\eta SE}^{\chi\text{-rec}})) \subseteq \mathrm{Wff}(\mathrm{AR}_E^{\text{rec}}).$$

Alternatively, let $\mathfrak{u} \in \mathrm{Wff}(\mathrm{AR}_E^{\text{rec}})$. By Proposition 10 of §4.15 we have $B_{\mathfrak{u}} \neq \emptyset$, in which case by (1) above for some $\mathfrak{w} \in \mathrm{Wff}_S^{\text{total}}$ we have $\overline{\eta}(\mathfrak{w}) = \mathfrak{u}$.

Now for some $n \geq 0$,

(3) <u>let</u> $g_1, g_2, \ldots, g_n$ <u>be the functions (not necessarily distinct) occurring from left to right in</u> $\mathfrak{u}$. <u>(That is, let</u> g_1 <u>be the leftmost function occurring in</u> $\mathfrak{u}$, <u>let</u> g_2 <u>be the second-from-left function occurring in</u> $\mathfrak{u}$, <u>and so on.)</u>

With $\mathfrak{u} \in \mathrm{Wff}(\mathrm{AR}_E^{\text{rec}})$ we obviously (cf. §§4.2 and 4.4) have

$$g_1, g_2, \ldots, g_n \in \mathrm{F}_E^{\text{rec}}.$$

Then by Proposition 7 of §4.15

(4) there exist unique functions

$$f_1, f_2, \ldots, f_n \in \mathrm{F}_{\eta SE}^{\text{scar-}\chi\text{-rec}}$$

(in which case, by Definition 5 of §4.15,

$$f_i = \chi(g_i) \quad (i = 1, 2, \ldots, n),$$

so that (with $\chi : \mathrm{F}_E^{\text{rec}} \longrightarrow \mathrm{F}_S^{\text{total}}$) we have

$$f_1, f_2, \ldots, f_n \in \mathrm{F}_S^{\text{total}})$$

such that $\eta[f_i] = g_i$ $(i = 1, 2, \ldots, n)$.

(5) For any symbol s used in constructing wffs of $\mathrm{Wff}_S^{\text{total}}$, we define $\eta(s)$ as follows (cf. Remark 1 at the end of this section):

$$\eta(s) = \begin{cases} \eta(s) & \text{if } s \text{ is a constant term,} \\ \eta[s] & \text{if } s \text{ is a function,} \\ s & \text{otherwise (i.e., if } s \text{ is a logical connective, parenthesis, numerical variable, etc.).} \end{cases}$$

We see (§4.15, Definition 2) that

(6) for any wff $\varpi \in \mathrm{Wff}_S^{\text{total}}$, if ϖ is a string of symbols $s_1 s_2 \ldots s_t$, for some $t \geq 1$, then

$$\overline{\eta}(\varpi) = \eta(s_1)\eta(s_2)\ldots\eta(s_t).$$

Now for some $t \geq 1$ and for not necessarily distinct symbols s_1, s_2, ..., s_t, let $\mathfrak{w} = s_1 s_2 \ldots s_t$. Then, with $\overline{\eta}(\mathfrak{w}) = \mathfrak{u}$, by (6) we have

(7) $$\overline{\eta}(\mathfrak{w}) = \eta(s_1)\eta(s_2)\ldots\eta(s_t) = \mathfrak{u}.$$

From s_1, s_2, ..., s_t being symbols it follows by (5) that $\eta(s_1)$, $\eta(s_2)$, ..., $\eta(s_t)$ are also symbols. Then from (3) and (7) it follows that

(8) <u>for some</u> i_1, i_2, ..., i_n, <u>with</u>

$$1 \leq i_1 < i_2 < \ldots < i_n \leq t,$$

<u>we have</u> $\eta(s_{i_1}) = g_1$, $\eta(s_{i_2}) = g_2$, ..., $\eta(s_{i_n}) = g_n$.

From this, with g_1, g_2, ..., g_n being functions it follows that $\eta(s_{i_1})$, $\eta(s_{i_2})$, ..., $\eta(s_{i_n})$ are also functions, and with (5) we see that s_{i_1}, s_{i_2}, ..., s_{i_n} are functions, and, in fact, are the only functions in $\mathfrak{w}$. (For, by (3), g_1, g_2, ..., g_n are exactly those functions (from left to right) in $\mathfrak{u}$, and (by (5)) η maps functions (and only functions) to functions, and by (8) we see that s_{i_1}, s_{i_2}, ..., s_{i_n} are exactly those functions in $\mathfrak{w}$ (where s_{i_1}, s_{i_2}, ..., s_{i_n} are not necessarily distinct).)

Recalling that $\mathfrak{w} = s_1 s_2 \ldots s_t$, we form a formula $\mathfrak{w}_0$ by replacing s_{i_1}, s_{i_2}, ..., s_{i_n} by f_1, f_2, ..., f_n, respectively. From $\mathfrak{w} \in \mathrm{Wff}_S^{\mathrm{total}}$, with (4) it follows by Definition 1α of §4.15 that we also have $\mathfrak{w}_0 \in \mathrm{Wff}_S^{\mathrm{total}}$.

By the preceding two paragraphs we see that f_1, f_2, ..., f_n are exactly those functions occurring in $\mathfrak{w}_0$ (where f_1, f_2, ..., f_n need not be distinct).

By the construction of $\mathfrak{w}_0$ we have

$$\mathfrak{w}_0 = s_1 \dots s_{i_1-1} f_1 s_{i_1+1} \dots s_{i_2-1} f_2 s_{i_2+1} \dots s_{i_n-1} f_n s_{i_n+1} \dots s_t,$$

so that by (6) and (8) we have

$$\begin{aligned}\bar{\eta}(\mathfrak{w}_0) = \eta(s_1) \dots \eta(s_{i_1-1}) g_1 \eta(s_{i_1+1}) \dots \\ \eta(s_{i_2-1}) g_2 \eta(s_{i_2+1}) \dots \\ \eta(s_{i_n-1}) g_n \eta(s_{i_n+1}) \dots \eta(s_t),\end{aligned}$$

and since (by (7)) $\bar{\eta}(\mathfrak{w}) = \eta(s_1)\eta(s_2)\dots\eta(s_t) = \mathfrak{u}$, with (8) we clearly have

$$\bar{\eta}(\mathfrak{w}_0) = \bar{\eta}(\mathfrak{w}) = \mathfrak{u}.$$

By part (b) of Definition 7 of §4.15, from

$$\bar{\eta}(\mathfrak{w}_0) = \mathfrak{u} \in \mathrm{Wff}(\mathrm{AR}_E^{\mathrm{rec}})$$

it follows that $\mathfrak{w}_0 \in \mathfrak{W}_{\eta SE}^{\mathrm{scar\text{-}rec}}$ and that $\mathfrak{w}_0$ is a recursive repetition of $\mathfrak{u}$.

With $f_1, f_2, \dots, f_n \in \mathrm{F}_{\eta SE}^{\mathrm{scar\text{-}\chi\text{-}rec}}$ (in (4)) and f_1, f_2, ..., f_n being the only functions occurring in $\mathfrak{w}_0$, it follows by Proposition 14 of §4.15 that

$$\mathfrak{w}_0 \in \mathrm{Wff}(\mathrm{AR}_{\eta SE}^{\mathrm{scar\text{-}\chi\text{-}rec}}),$$

or briefly (§4.15, Definition 9, part (III))

$$\mathfrak{w}_0 \in \mathrm{Wff}(\mathrm{AR}^{\chi\text{-rec}}_{\eta SE}).$$

From this, with $\overline{\eta}(\mathfrak{w}_0) = \mathfrak{u}$ we have $\mathfrak{u} \in \overline{\eta}(\mathrm{Wff}(\mathrm{AR}^{\chi\text{-rec}}_{\eta SE}))$.

Thus we have proved that

$$\mathrm{Wff}(\mathrm{AR}^{\mathrm{rec}}_{E}) \subseteq \overline{\eta}(\mathrm{Wff}(\mathrm{AR}^{\chi\text{-rec}}_{\eta SE})),$$

and the proposition follows.

We now prove four lemmas for Theorems 1 and 2 of the following section (those theorems being placed in a separate section because of the length of their proofs):

$\mathscr{F}_{BS}$-LEMMA 1 (for Theorems 1 and 2 of §4.20). Let S be an atomic infinite-dimensional nonzeroed semiological space, let ηSE be a Peano semicarrier space, let $\overline{\eta}$ be the canonical recursive-extension of η to wffs, and let χ: $\mathrm{F}^{\mathrm{rec}}_{E} \longrightarrow \mathrm{F}^{\mathrm{total}}_{S}$ be a total function of the type specified in Definition 5 of §4.15. Then:

(i) If $\mathfrak{w}_0$ is an axiom of $\mathrm{AR}^{\chi\text{-rec}}_{\eta SE}$, then $\overline{\eta}(\mathfrak{w}_0)$ is an axiom of $\mathrm{AR}^{\mathrm{rec}}_{E}$.

(ii) If $\mathfrak{u}$ is an axiom of $\mathrm{AR}^{\mathrm{rec}}_{E}$, then there exists an axiom $\mathfrak{w}_0$ of $\mathrm{AR}^{\chi\text{-rec}}_{\eta SE}$ such that $\overline{\eta}(\mathfrak{w}_0) = \mathfrak{u}$.

Proof. First assume that $\mathfrak{w}_0$ is an axiom of $\mathrm{AR}^{\chi\text{-rec}}_{\eta SE}$. Then by condition (IV) of Definition 9 of §4.15, $\overline{\eta}(\mathfrak{w}_0)$ is an axiom of $\mathrm{AR}^{\mathrm{rec}}_{E}$, and part (i) of the lemma is established.

Alternatively, assume that $\mathfrak{u}$ is an axiom of $\mathrm{AR}^{\mathrm{rec}}_{E}$, in which case $\mathfrak{u} \in \mathrm{Wff}(\mathrm{AR}^{\mathrm{rec}}_{E})$. Then by Proposition 19 of §4.15

there exists a $\mathfrak{w}_0 \in \mathrm{Wff}(\mathrm{AR}_{\eta SE}^{\mathrm{scar}\text{-}\chi\text{-}\mathrm{rec}})$ such that $\overline{\eta}(\mathfrak{w}_0) = \mathfrak{u}$. With this and $\mathfrak{u}$ being an axiom of $\mathrm{AR}_E^{\mathrm{rec}}$ it follows that $\overline{\eta}(\mathfrak{w}_0)$ is an axiom of $\mathrm{AR}_E^{\mathrm{rec}}$. By condition (III) of Definition 9 of §4.15 it follows that $\mathfrak{w}_0$ is a wff of $\mathrm{AR}_{\eta SE}^{\chi\text{-}\mathrm{rec}}$. Then by condition (IV) of Definition 9 of §4.15, $\mathfrak{w}_0$ is an axiom of $\mathrm{AR}_{\eta SE}^{\chi\text{-}\mathrm{rec}}$, and part (ii) of the lemma is established.

$\mathscr{F}_{\mathrm{BS}}$**-LEMMA 2** (for Theorems 1 and 2 of §4.20). Let S be an atomic infinite-dimensional nonzeroed semiological space, let ηSE be a Peano semicarrier space, let $\overline{\eta}$ be the canonical recursive-extension of η to wffs, and let $\chi\colon \mathrm{F}_E^{\mathrm{rec}} \longrightarrow \mathrm{F}_S^{\mathrm{total}}$ be a total function of the type specified in Definition 5 of §4.15. Then, along the lines of (5) and (6) in the proof of Proposition 1 of the present section, we extend η to terms, functions, and relations (considered as strings of symbols) in the obvious way and have the following results:

(i) If T is a term of $\mathrm{AR}_{\eta SE}^{\chi\text{-}\mathrm{rec}}$, then $\overline{\eta}(T)$ is a term of $\mathrm{AR}_E^{\mathrm{rec}}$.

(ii) If U is a term of $\mathrm{AR}_E^{\mathrm{rec}}$, then there exists a term T of $\mathrm{AR}_{\eta SE}^{\chi\text{-}\mathrm{rec}}$ such that $\overline{\eta}(T) = U$.

(iii) For any $n \geq 1$, if Δ is an n-place function (resp. relation) of $\mathrm{AR}_{\eta SE}^{\chi\text{-}\mathrm{rec}}$ and $T_1, T_2, \ldots, T_n$ are terms of $\mathrm{AR}_{\eta SE}^{\chi\text{-}\mathrm{rec}}$, then $\overline{\eta}(\Delta)$ is an n-place function (resp. relation) of $\mathrm{AR}_E^{\mathrm{rec}}$ and we have

$$\overline{\eta}(\Delta(T_1,T_2,\ldots,T_n)) = \overline{\eta}(\Delta)(\overline{\eta}(T_1),\overline{\eta}(T_2),\ldots,\overline{\eta}(T_n)).$$

(iv) For any $n \geq 1$, if Δ_0 is an n-place function (resp. relation) of $\mathrm{AR}_E^{\mathrm{rec}}$ and U_1, U_2, ..., U_n are terms in $\mathrm{AR}_E^{\mathrm{rec}}$, then there exist an n-place function (resp. relation) Δ of $\mathrm{AR}_{\eta SE}^{\chi\text{-rec}}$ and terms T_1, T_2, ..., T_n of $\mathrm{AR}_{\eta SE}^{\chi\text{-rec}}$ such that $\overline{\eta}(\Delta) = \Delta_0$ and

$$\overline{\eta}(T_i) = U_i \quad (i = 1, 2, \ldots, n)$$

and such that

$$\overline{\eta}(\Delta(T_1,T_2,\ldots,T_n)) = \overline{\eta}(\Delta)(\overline{\eta}(T_1),\overline{\eta}(T_2),\ldots,\overline{\eta}(T_n)).$$

Proof. If a string of symbols $s_1s_2\ldots s_n$ $(n \geq 1)$ is a term (resp. a function, resp. a relation) of $\mathrm{AR}_{\eta SE}^{\chi\text{-rec}}$, then by (5) in the proof of Proposition 1 of the present section it easily follows that the formula

$$\overline{\eta}(s_1s_2\ldots s_n)$$

$(= \eta(s_1)\eta(s_2)\ldots\eta(s_n))$ is a term (resp. a function, resp. a relation) of $\mathrm{AR}_E^{\mathrm{rec}}$. With this, one easily establishes parts (i) and (iii) of the lemma.

Alternatively, let $s'_1s'_2\ldots s'_n$ $(n \geq 1)$ be a string of symbols composing a term U (resp. a function $f_0(x)$, resp. a relation $R_0(x)$) of $\mathrm{AR}_E^{\mathrm{rec}}$. We need to construct an expression $s_1s_2\ldots s_n$. Letting $1 \leq i \leq n$, we determine the symbol s_i:

(a) If s'_i is a constant term b of E, then from $\eta: S \longrightarrow E$ being a surjection (§4.15, Definition 1) it follows that there exists a constant term a of S such that $\eta(a) = b$ (= s'_i), and we set $s_i = a$.

(b) Assume that s'_i is a function g of $\mathbf{F}_E^{rec}$. By Proposition 7 of §4.15 and Definition 5 of §4.15 there exists a unique function $f \in \mathbf{F}_{\eta SE}^{scar-\chi-rec}$ such that $\chi(g) = f$ and $\eta[f] = g$, and we set $s_i = f$.

(c) If s'_i is neither a constant term nor a function, then we set $s_i = s'_i$ and by (5) in the proof of Proposition 1 of the present section we see that $\eta(s_i) = s'_i$.

From the preceding we see that the expression $s_1 s_2 \ldots s_n$ is a term T (resp. a function $f(x)$, resp. a relation $R(x)$) of arithmetic $\mathrm{AR}_{\eta SE}^{\chi-rec}$ and that $\overline{\eta}(T)$ (resp. $\overline{\eta}(f(x))$, resp. $\overline{\eta}(R(x))$) satisfies:

$$\begin{aligned}\overline{\eta}(T) \text{ (resp. } \overline{\eta}(f(x)) \ (= \overline{\eta}(f)(x)),& \\ \text{resp. } \overline{\eta}(R(x)) \ (= \overline{\eta}(R)(x))) &= \overline{\eta}(s_1 s_2 \ldots s_n) \\ &= \eta(s_1)\eta(s_2)\ldots\eta(s_n) \\ &= s'_1 s'_2 \ldots s'_n \\ &= U \text{ (resp. } = f_0(x), \\ &\qquad \text{resp. } = R_0(x)).\end{aligned}$$

(Our having $\overline{\eta}(f(x)) = \overline{\eta}(f)(x)$ and $\overline{\eta}(R(x)) = \overline{\eta}(R)(x)$ follows from $\eta(x) = x$ in (5) in the proof of Proposition 1 of the present section.)

One then easily proves that parts (ii) and (iv) of the

lemma hold.

Note, the following lemma deals with _sentential_ arithmetics:

$\mathscr{F}_{BS}$-LEMMA 3 (for Theorems 1 and 3 of §4.20). _Let S be an atomic infinite-dimensional nonzeroed semiological space, let ηSE be a Peano semicarrier space, let $\bar{\eta}$ be the canonical recursive-extension of η to wffs, and let χ:$F_E^{rec} \longrightarrow F_S^{total}$ be a total function of the type specified in_ Definition 5 _of_ §4.15. _Then:_

(i) _If $\mathfrak{w}_0$ is a wff of_ $SAR_{\eta SE}^{\chi\text{-}rec}$ _that is an instance of one of axiom schemas_ S1, S2, S3, S4, S6 _of_ Bourbaki, _then $\bar{\eta}(\mathfrak{w}_0)$ is a wff of_ SAR_E^{rec} _that is an instance of the same axiom schema._

(ii) _If $\mathfrak{u}$ is a wff of_ SAR_E^{rec} _that is an instance of one of axiom schemas_ S1, S2, S3, S4, S6 _of_ Bourbaki, _then there exists a wff $\mathfrak{w}_0$ of_ $SAR_{\eta SE}^{\chi\text{-}rec}$ _such that $\bar{\eta}(\mathfrak{w}_0) = \mathfrak{u}$ and such that $\mathfrak{w}_0$ is an instance of the same axiom schema._

Proof. Part (i) of the lemma follows immediately by Proposition 1α of §4.15 and Lemma 2 for Theorems 1 and 2 of §4.20 (that lemma being given above in the present section).

Alternatively, assume that $\mathfrak{u}$ is a wff of SAR_E^{rec} that is an instance of one of axiom schemas S1, S2, S3, S4, S6 of Bourbaki. By Proposition 19 of §4.15 there exists a

$$\mathfrak{w}_0 \in \mathrm{Wff}(SAR_{\eta SE}^{scar\text{-}\chi\text{-}rec})$$

such that $\bar{\eta}(\mathfrak{w}_0) = \mathfrak{u}$. By condition (III) of Definition 9 of

§4.15, $\mathfrak{w}_0$ is a wff of $SAR_{\eta SE}^{\chi\text{-rec}}$. Moreover, by Proposition 1α of §4.15 and Lemma 2 for Theorems 1 and 2 of §4.20 (that lemma being given above in the present section) we see that $\mathfrak{w}_0$ is an instance of the same axiom schema as $\mathfrak{u}$, and part (ii) of the lemma is established.

Note, the following lemma deals with <u>equational</u> arithmetics (note, the Axiom Schemas I and II referred to are in Definition 9 of §4.15):

$\mathcal{F}_{BS}$**-LEMMA 4** (for Theorems 1 and 2 of §4.20). <u>Let</u> S <u>be an atomic infinite-dimensional nonzeroed semiological space, let</u> ηSE <u>be a Peano semicarrier space, let</u> $\bar{\eta}$ <u>be the canonical recursive-extension of</u> η <u>to wffs, and let</u> $\chi : F_E^{rec} \longrightarrow F_S^{total}$ <u>be a total function of the type specified in</u> Definition 5 <u>of</u> §4.15. <u>Then</u>:

(i) <u>if</u> $\mathfrak{w}_0$ <u>is a wff of</u> $EAR_{\eta SE}^{\chi\text{-rec}}$ <u>that is an instance of</u> Axiom Schema I (resp. Axiom Schema II) <u>of</u> $EAR_{\eta SE}^{\chi\text{-rec}}$, <u>then</u> $\bar{\eta}(\mathfrak{w}_0)$ <u>is a wff of</u> EAR_E^{rec} <u>that is an instance of</u> Axiom Schema I (resp. Axiom Schema II) <u>of</u> EAR_E^{rec};

(ii) <u>if</u> $\mathfrak{w}$ <u>is a wff of</u> EAR_E^{rec} <u>that is an instance of</u> Axiom Schema I (resp. Axiom Schema II) <u>of</u> EAR_E^{rec}, <u>then there exists a wff</u> $\mathfrak{w}_0$ <u>of</u> $EAR_{\eta SE}^{\chi\text{-rec}}$ <u>such that</u> $\bar{\eta}(\mathfrak{w}_0) = \mathfrak{w}$ <u>and such that</u> $\mathfrak{w}_0$ <u>is an instance of</u> Axiom Schema I (resp. Axiom Schema II) <u>of</u> $EAR_{\eta SE}^{\chi\text{-rec}}$.

<u>Proof</u>. One proves part (i) of the lemma in a straightforward manner, noting that the axiom schemas of $EAR_{\eta SE}^{\chi\text{-rec}}$ are Axiom Schemas I and II of Definition 9 of §4.15 and that the axiom

schemas of EAR_E^{rec} are Axiom Schemas I and II of §4.4. One uses the fact that for all $\mathbf{g} \in \mathrm{F}_E^{rec}$ we have $\eta[\chi(\mathbf{g})] = \mathbf{g}$ (cf. Definition 5 of §4.15). One also uses relations (5) and (6) of the proof of Proposition 1 of the present section.

We now establish part (ii) of the lemma in the case of Axiom Schema II (the case of Axiom Schema I is handled along the same lines, but more simply). Assume $\mathfrak{w}$ is an instance of Axiom Schema II of EAR_E^{rec}, say

$$\mathfrak{w} = (1 \dot{-} (\sigma_E(z) \dot{-} f_i^n(x_1, x_2, \ldots, x_n))) \times \\ (1 \dot{-} \gamma_E(x_1, x_2, \ldots, x_n, z)) = 0,$$

where $n \geq 1$, $i \geq 0$, and $\gamma_E \in \mathrm{F}_{\eta SE}^{scar-\chi-rec}$. We clearly have $\mathfrak{w} \in \mathrm{Wff}(\mathrm{EAR}_E^{rec})$ and by Proposition 19 of §4.15 there exists a $\mathfrak{u} \in \mathrm{Wff}(\mathrm{EAR}_{\eta SE}^{scar-\chi-rec})$ such that $\overline{\eta}(\mathfrak{u}) = \mathfrak{w}$. By condition (III) of Definition 9 of §4.15 we have $\mathfrak{u} \in \mathrm{Wff}(\mathrm{EAR}_{\eta SE}^{\chi-rec})$.

Now let $\mathfrak{u}$ be the string of symbols $\mathfrak{u} = s_1 s_2 \ldots s_t$ $(t \geq 1)$. By (5) and (6) in the proof of Proposition 1 of the present section we have

$$\overline{\eta}(\mathfrak{u}) = \eta(s_1)\eta(s_2)\ldots\eta(s_t),$$

which with $\overline{\eta}(\mathfrak{u}) = \mathfrak{w}$ yields

$$\eta(s_1)\eta(s_2)\ldots\eta(s_t) = \\ (1 \dot{-} (\sigma_E(z) \dot{-} f_i^n(x_1, x_2, \ldots, x_n))) \times \\ (1 \dot{-} \gamma_E(x_1, x_2, \ldots, x_n, z)) = 0.$$

From this it follows by (5) and (6) in the proof of Proposition 1 of the present section that

$$\mathfrak{u} = s_1 s_2 \ldots s_t = (c\ \Delta_1\ (g(z)\ \Delta_1\ h(x_1, x_2, \ldots, x_n)))\ \Delta_2\ (c\ \Delta_1\ h'(x_1, x_2, \ldots, x_n, z)) = d,$$

where c and d are constants of S (with $\eta(c) = 1$ and $\eta(d) = 0$) and where Δ_1, g, h, Δ_2, h' are S-functions (with $\eta[\Delta_1] = \dot{-}$, $\eta[g] = \sigma_E$, $\eta[h] = f^n_i$, $\eta[\Delta_2] = \times$, and $\eta[h'] = \gamma_E$).

From $\chi\colon \mathbf{F}_E^{\text{rec}} \longrightarrow \mathbf{F}_S^{\text{total}}$ being total it follows that each of $\chi(\dot{-})$, $\chi(\sigma_E)$, $\chi(f^n_i)$, $\chi(\times)$, and $\chi(\gamma_E)$ is defined. Moreover, by Definition 5 of §4.15 we have

$$\chi(\dot{-}),\ \chi(\sigma_E),\ \chi(f^n_i),\ \chi(\times),\ \chi(\gamma_E) \in \mathbf{F}_{\eta SE}^{\text{scar-}\chi\text{-rec}}$$

and (by the same reference)

$$\eta[\chi(\dot{-})] = \dot{-},\ \eta[\chi(\sigma_E)] = \sigma_E,\ \eta[\chi(f^n_i)] = f^n_i,$$
$$\eta[\chi(\times)] = \times,\ \eta[\chi(\gamma_E)] = \gamma_E.$$

Also by Definition 5 of §4.15 we see that $\chi(\dot{-})$, $\chi(\sigma_E)$, $\chi(f^n_i)$, $\chi(\times)$, and $\chi(\gamma_E)$ are the <u>only</u> functions in $\mathbf{F}_{\eta SE}^{\text{scar-}\chi\text{-rec}}$ that are mapped by η to $\dot{-}$, σ_E, f^n_i, $\times$, and γ_E, respectively. So we have $\Delta_1 = \chi(\dot{-})$, $g = \chi(\sigma_E)$, $h = \chi(f^n_i)$, $\Delta_2 = \chi(\times)$, $h' = \chi(\gamma_E)$.

With $\eta(c) = 1$, $\eta(d) = 0$, and ηSE being a Peano semicarrier space we obtain

$$\eta(\#_S) = \#_E\ (= 0 = \eta(d))$$

(§4.15, Definition 1) and hence also

$$\eta(\chi(\sigma_E)(\#_S)) = \eta[\chi(\sigma_E)](\eta(\#_S)) = \sigma_E(0) = 1 = \eta(c)$$

(by (5) and (6) in the proof of Proposition 1 of the present section).

So we define a wff $\mathfrak{w}_0$ by replacing each occurrence of c by $\chi(\sigma_E)(\#_S)$ and replacing d by $\#_S$ in $\mathfrak{u}$, and we see that

$$\overline{\eta}(\mathfrak{w}_0) = \overline{\eta}(\mathfrak{u}) = \mathfrak{w}.$$

With $\mathfrak{u}$ being a wff of $\mathrm{EAR}^{\chi\text{-rec}}_{\eta SE}$, we see that $\mathfrak{w}_0$ is a wff of $\mathrm{EAR}^{\chi\text{-rec}}_{\eta SE}$ and we have

$$\begin{aligned}\mathfrak{w}_0 = (\chi(\sigma_E)(\#_S)\ \chi(\dot{-})\ (\chi(\sigma_E)(z)\ \chi(\dot{-})\\ \chi(f^n_i)(x_1,x_2,\ldots,x_n)))\ \chi(\times)\\ (\chi(\sigma_E)(\#_S)\ \chi(\dot{-})\ \chi(\gamma_E)(x_1,x_2,\ldots,x_n,z)) = \#_S,\end{aligned}$$

which we see is an instance of Axiom Schema II of item (V) of Definition 9 of §4.15.

As mentioned above, the theorems for which the preceding four lemmas are proved are given in the following section and, in fact, are the sole items of that section.

We conclude this section with a brief remark reconciling (i) the fact that our arithmetics have been defined with the primitive symbols not being specified and (ii) our applying the mapping η to symbols (cf. relations (5) and (6) in the proof of Proposition 1):

REMARK 1. We recall (cf. §4.15) that the wffs of our semi-carrier arithmetics are not defined with complete formality, i.e., beginning with lists of primitive symbols. Likewise (cf. §4.4), the wffs of our equational arithmetic EAR_{E}^{rec} are also defined without listing the primitive symbols. Therefore, in order to be able to talk meaningfully of applying the semi-carrier map η to symbols (cf. relations (5) and (6) in the proof of Proposition 1) we establish the following conventions:

(i) In each of the arithmetics under consideration (a) we will consider each function in a wff (where the function in question is considered only as a single function) to be denoted by a single symbol and (b) we will consider each constant term in a wff (where the term in question is considered only as a single term) to be denoted by a single symbol.

(ii) In each of the arithmetics under consideration (a) we will consider each composition of functions in a wff (where the composition of functions is considered only as a composition, and not as a single function) ([1]) to be denoted not as a single symbol, but rather in the usual way as a string of symbols consisting of the symbols denoting the functions being composed and the composition sign "∘", and (b) we will consider each resolution $\sigma_1 \circ \ldots \circ \sigma_\nu(t)$ $(\nu \geq 1)$ in a wff (where the resolution

---

([1]) Although a composition of functions is, in fact, itself a single function.

is considered only as a resolution, and not as a value) [2] to be denoted not as a single symbol, but rather as a string of symbols consisting of the symbols denoting the functions in the resolution, the composition sign "∘", the parentheses, and the string of symbols denoting the term **t**.

---

(2) Even though a resolution can be considered as a value.

§4.20. We devote this section to proving two theorems which assert, respectively, that, given a semicarrier recursive arithmetic $AR^{\chi\text{-rec}}_{\eta SE}$, the extension $\bar{\eta}$ of η to wffs maps every proof in $AR^{\chi\text{-rec}}_{\eta SE}$ to a proof in AR^{rec}_{E} and, conversely, every proof in AR^{rec}_{E} is the image under $\bar{\eta}$ of a proof in $AR^{\chi\text{-rec}}_{\eta SE}$.

$\mathscr{F}_{BS}$-THEOREM 1. *Let S be an atomic infinite-dimensional non-zeroed semiological space, let ηSE be a Peano semicarrier space, let $\bar{\eta}$ be the canonical recursive-extension of η to wffs, and let $\chi\colon F^{\text{rec}}_{E} \longrightarrow F^{\text{total}}_{S}$ be a total function of the type specified in* Definition 5 *of* §4.15. *Then, for each $n \geq 1$, if*

$$\mathfrak{w}_1,\ \mathfrak{w}_2,\ \ldots,\ \mathfrak{w}_n$$

is a proof in $AR^{\chi\text{-rec}}_{\eta SE}$, *then*

$$\bar{\eta}(\mathfrak{w}_1),\ \bar{\eta}(\mathfrak{w}_2),\ \ldots,\ \bar{\eta}(\mathfrak{w}_n)$$

is a proof in AR^{rec}_{E}.

Proof. We prove the theorem using induction on n. First, if $\mathfrak{w}_1$ is a proof in $AR^{\chi\text{-rec}}_{\eta SE}$, then $\mathfrak{w}_1$ is either an explicit axiom or instance of an axiom schema of $AR^{\chi\text{-rec}}_{\eta SE}$, and by Lemmas 1, 3, and 4 (which are in §4.19) it follows that $\bar{\eta}(\mathfrak{w}_1)$ is likewise either an explicit axiom or instance of an axiom schema of AR^{rec}_{E}, and hence is a proof in AR^{rec}_{E}.

For some $n \geq 1$, assume as induction hypothesis that the theorem holds for proofs in $AR^{\chi\text{-rec}}_{\eta SE}$ of length n, and let $\mathfrak{w}_1$, $\mathfrak{w}_2$, ..., $\mathfrak{w}_n$, $\mathfrak{w}_{n+1}$ be a proof in $AR^{\chi\text{-rec}}_{\eta SE}$. Then $\mathfrak{w}_1$, $\mathfrak{w}_2$, ..., $\mathfrak{w}_n$ is also a proof in $AR^{\chi\text{-rec}}_{\eta SE}$, and by the induction hypothesis it follows that $\overline{\eta}(\mathfrak{w}_1)$, $\overline{\eta}(\mathfrak{w}_2)$, ..., $\overline{\eta}(\mathfrak{w}_n)$ is a proof in AR_E^{rec}.

First, if $\mathfrak{w}_{n+1}$ is either an explicit axiom or instance of an axiom schema of $AR_{\eta SE}^{\chi\text{-rec}}$, then by Lemmas 1, 3, and 4 (which are in §4.19) it follows that $\overline{\eta}(\mathfrak{w}_{n+1})$ is likewise either an explicit axiom or instance of an axiom schema of AR_E^{rec}, in which case $\overline{\eta}(\mathfrak{w}_1)$, $\overline{\eta}(\mathfrak{w}_2)$, ..., $\overline{\eta}(\mathfrak{w}_n)$, $\overline{\eta}(\mathfrak{w}_{n+1})$ is a proof in AR_E^{rec}.

Alternatively, assume that $\mathfrak{w}_{n+1}$ is obtained from certain of the wffs $\mathfrak{w}_1$, $\mathfrak{w}_2$, ..., $\mathfrak{w}_n$ using an inference rule of $AR_{\eta SE}^{\chi\text{-rec}}$. We have two cases to consider:

The Sentential Case. Assume that we are dealing with sentential arithmetics, in which case instead of AR we may write SAR. Then by part (VI) of Definition 9 of §4.15 we see that the inference rules of $SAR_{\eta SE}^{\chi\text{-rec}}$ consist of the inference rules of SAR_E^{rec} (applied to wffs of $SAR_{\eta SE}^{\chi\text{-rec}}$), where the successor function σ_E is replaced by the function $\chi(\sigma_E)$, which (see part (XI) of §4.2) implies that there are three inference rules that we have to consider:

<u>First</u>, assume that for some k, $1 \le k \le n$, and some ℓ, $1 \le \ell \le n$, the wff $\mathfrak{w}_{n+1}$ is obtained from $\mathfrak{w}_k$ and $\mathfrak{w}_\ell$ by modus ponens, in which case we may take $\mathfrak{w}_k$ to be the wff $\mathfrak{w}_\ell \Rightarrow \mathfrak{w}_{n+1}$. By Proposition 1α of §4.15 we have

$$\overline{\eta}(\mathfrak{w}_k) = \overline{\eta}(\mathfrak{w}_\ell \Rightarrow \mathfrak{w}_{n+1}) = [\overline{\eta}(\mathfrak{w}_\ell) \Rightarrow \overline{\eta}(\mathfrak{w}_{n+1})],$$

and from this it follows that $\overline{\eta}(\mathfrak{w}_{n+1})$ can be obtained from $\overline{\eta}(\mathfrak{w}_k)$ (= $\overline{\eta}(\mathfrak{w}_\ell) \Rightarrow \overline{\eta}(\mathfrak{w}_{n+1})$) and $\overline{\eta}(\mathfrak{w}_\ell)$ by modus ponens, which is an inference rule of $\mathrm{SAR}_E^{\mathrm{rec}}$ (see part (XI) of §4.2).

Second, assume that for some k, $1 \le k \le n$, the wff $\mathfrak{w}_{n+1}$ is obtained from $\mathfrak{w}_k$ by substitution of a term T for a variable x. So we can let $\mathfrak{w}_k$ be a relation $R(x)$ of $\mathrm{SAR}_{\eta SE}^{\chi\text{-rec}}$ and we have

$$\mathfrak{w}_{n+1} = (T|x)\mathfrak{w}_k = R(T),$$

so that by part (iii) of Lemma 2 (which is §4.19) we have

$$\begin{aligned}\bar{\eta}(\mathfrak{w}_{n+1}) &= \bar{\eta}(R(T)) \\ &= \bar{\eta}(R)(\bar{\eta}(T)) \\ &= (\bar{\eta}(T)|x)\bar{\eta}(R(x)) \\ &= (\bar{\eta}(T)|x)\bar{\eta}(\mathfrak{w}_k).\end{aligned}$$

Thus $\bar{\eta}(\mathfrak{w}_{n+1})$ can be obtained from $\bar{\eta}(\mathfrak{w}_k)$ by substitution of a term for a variable, which is an inference rule of $\mathrm{SAR}_E^{\mathrm{rec}}$ (part (XI) of §4.2).

Third, assume that for some k, $1 \le k \le n$, and some ℓ, $1 \le \ell \le n$, the wff $\mathfrak{w}_{n+1}$ is obtained from $\mathfrak{w}_k$ and $\mathfrak{w}_\ell$ by induction. Then $\mathfrak{w}_{n+1}$ is some relation $R(x)$ of $\mathrm{SAR}_{\eta SE}^{\chi\text{-rec}}$, and $\mathfrak{w}_k$ and $\mathfrak{w}_\ell$ are the relations $R(\#_S)$ and

$$R(x) \Rightarrow R(\chi(\sigma_E)(x)),$$

respectively (recall that $\chi(\sigma_E)$ is, in effect, the successor function of arithmetic $\mathrm{SAR}_{\eta SE}^{\chi\text{-rec}}$). By Lemma 2 (which is in §4.19) we see that

$$\bar{\eta}(\mathfrak{w}_k) = \bar{\eta}(R(\#_S)) = \bar{\eta}(R)(\eta(\#_S)) = \bar{\eta}(R)(0)$$

(we have $\eta(\#_S) = 0$ by Definition 1 of §4.15), and by Proposition 1α of §4.15 we have

$$\begin{aligned}\bar{\eta}(\mathfrak{w}_\ell) &= \bar{\eta}(R(x) \Rightarrow R(\chi(\sigma_E)(x)))\\ &= \bar{\eta}(R(x)) \Rightarrow \bar{\eta}(R(\chi(\sigma_E)(x)))\\ &= \bar{\eta}(R)(x) \Rightarrow \bar{\eta}(R)(\eta(\chi(\sigma_E)(x)))\\ &= \bar{\eta}(R)(x) \Rightarrow \bar{\eta}(R)(\eta[\chi(\sigma_E)](x)))\\ &= \bar{\eta}(R)(x) \Rightarrow \bar{\eta}(R)(\sigma_E(x))\end{aligned}$$

(recall that $\eta[\chi(\sigma_E)] = \sigma_E$ by Definition 5 of §4.15).

By part (iii) of Lemma 2 (which is in §4.19) we have

$$\bar{\eta}(\mathfrak{w}_{n+1}) = \bar{\eta}(R(x)) = \bar{\eta}(R)(x),$$

and then by item (2) in part (XI) of §4.2 we see that the wff $\bar{\eta}(\mathfrak{w}_{n+1}) = \bar{\eta}(R)(x)$ can be obtained from $\bar{\eta}(\mathfrak{w}_k)$ and $\bar{\eta}(\mathfrak{w}_\ell)$ by induction, which is an inference rule of $\mathrm{SAR}_E^{\mathrm{rec}}$ (see part (XI) of §4.2).

Thus in any event we see that in the sentential case the wff $\bar{\eta}(\mathfrak{w}_{n+1})$ can be obtained from certain of the wffs $\bar{\eta}(\mathfrak{w}_1)$, $\bar{\eta}(\mathfrak{w}_2)$, ..., $\bar{\eta}(\mathfrak{w}_n)$ by an inference rule of $\mathrm{SAR}_E^{\mathrm{rec}}$.

The Equational Case. Assume that we are dealing with equational arithmetics, in which case instead of AR we may write EAR. Then by part (VI) of Definition 9 of §4.15 we see that the inference rules of $\mathrm{EAR}_{\eta SE}^{\chi\text{-rec}}$ consist of the inference rules of EAR_E^{rec} (applied to wffs of $EAR_{\eta SE}^{\chi\text{-}rec}$), where the successor function σ_E is replaced by the function $\chi(\sigma_E)$, which (see part (XI) of §4.4) implies that there are six inference rules to consider:

<u>First</u>, assume that for some k, $1 \le k \le n$, the wff $\mathfrak{w}_{n+1}$ is obtained from $\mathfrak{w}_k$ by inference rule (E_1) (applied to wffs of $EAR_{\eta SE}^{\chi\text{-}rec}$). Then for terms $\mathbf{t}_1$ and $\mathbf{t}_2$ of $EAR_{\eta SE}^{\chi\text{-}rec}$ we have

$$\mathfrak{w}_k = [\mathbf{t}_1 = \mathbf{t}_2],$$

$$\mathfrak{w}_{n+1} = [\mathbf{t}_2 = \mathbf{t}_1],$$

and with Proposition 1α of §4.15 we have

$$\overline{\eta}(\mathfrak{w}_k) = \overline{\eta}(\mathbf{t}_1 = \mathbf{t}_2) = [\overline{\eta}(\mathbf{t}_1) = \overline{\eta}(\mathbf{t}_2)],$$

$$\overline{\eta}(\mathfrak{w}_{n+1}) = \overline{\eta}(\mathbf{t}_2 = \mathbf{t}_1) = [\overline{\eta}(\mathbf{t}_2) = \overline{\eta}(\mathbf{t}_1)],$$

and (see part (XI) of §4.4) we see that $\overline{\eta}(\mathfrak{w}_{n+1})$ can be obtained from $\overline{\eta}(\mathfrak{w}_k)$ by inference rule (E_1) of EAR_E^{rec}.

<u>Second</u>, assume that for some k, $1 \le k \le n$, and some ℓ, $1 \le \ell \le n$, the wff $\mathfrak{w}_{n+1}$ is obtained from $\mathfrak{w}_k$ and $\mathfrak{w}_\ell$ by inference rule (E_2) (applied to wffs of $EAR_{\eta SE}^{\chi\text{-}rec}$). Then for terms $\mathbf{t}_1$, $\mathbf{t}_2$, and $\mathbf{t}_3$ of $EAR_{\eta SE}^{\chi\text{-}rec}$ we have

$$\mathfrak{w}_k = [\mathbf{t}_1 = \mathbf{t}_2], \qquad \mathfrak{w}_\ell = [\mathbf{t}_2 = \mathbf{t}_3], \qquad \mathfrak{w}_{n+1} = [\mathbf{t}_1 = \mathbf{t}_3],$$

and with Proposition 1α of §4.15 we have

$$\overline{\eta}(\mathfrak{w}_k) = \overline{\eta}(\mathbf{t}_1 = \mathbf{t}_2) = [\overline{\eta}(\mathbf{t}_1) = \overline{\eta}(\mathbf{t}_2)],$$

$$\bar{\eta}(\mathfrak{w}_\ell) = \bar{\eta}(\mathbf{t}_2 = \mathbf{t}_3) = [\bar{\eta}(\mathbf{t}_1) = \bar{\eta}(\mathbf{t}_3)],$$

$$\bar{\eta}(\mathfrak{w}_{n+1}) = \bar{\eta}(\mathbf{t}_1 = \mathbf{t}_3) = [\bar{\eta}(\mathbf{t}_1) = \bar{\eta}(\mathbf{t}_3)],$$

and (see part (XI) of §4.4) we see that $\bar{\eta}(\mathfrak{w}_{n+1})$ can be obtained from $\bar{\eta}(\mathfrak{w}_k)$ and $\bar{\eta}(\mathfrak{w}_\ell)$ by inference rule (E_2) of $\text{EAR}_E^{\text{rec}}$.

<u>Third</u>, assume that for some k, $1 \le k \le n$, some ℓ, $1 \le \ell \le n$, and some m, $1 \le m \le n$, the wff $\mathfrak{w}_{n+1}$ is obtained from $\mathfrak{w}_k$, $\mathfrak{w}_\ell$, and $\mathfrak{w}_m$ by inference rule (IND) (applied to wffs of $\text{EAR}_{\eta SE}^{\chi\text{-rec}}$). Then for functions $f(x)$, $g(x)$, and $H(x,y)$ of $\text{EAR}_{\eta SE}^{\chi\text{-rec}}$ (where each of f, g, and H may have parameters in addition to x and y, with each of f, g, and H having exactly the same parameters) we have

$$\mathfrak{w}_k = [f(\#_S) = g(\#_S)],$$

$$\mathfrak{w}_\ell = [f(\chi(\sigma_E)(x)) = H(x,f(x))],$$

$$\mathfrak{w}_m = [g(\chi(\sigma_E)(x)) = H(x,g(x))],$$

$$\mathfrak{w}_{n+1} = [f(x) = g(x)].$$

From this, with Lemma 2 (which is in §4.19) and Proposition 1α of §4.15 we obtain

$$\begin{aligned}\bar{\eta}(\mathfrak{w}_k) &= \bar{\eta}(f(\#_S) = g(\#_S)) \\ &= [\bar{\eta}(f(\#_S)) = \bar{\eta}(g(\#_S))] \\ &= [\bar{\eta}(f)(\bar{\eta}(\#_S)) = \bar{\eta}(g)(\bar{\eta}(\#_S))] \\ &= [\bar{\eta}(f)(0) = \bar{\eta}(g)(0)],\end{aligned}$$

$$\bar{\eta}(\mathfrak{w}_{\ell}) = \bar{\eta}(f(\chi(\sigma_E)(x)) = H(x, f(x)))$$

$$= [\bar{\eta}(f(\chi(\sigma_E)(x))) = \bar{\eta}(H(x, f(x)))]$$

$$= [\bar{\eta}(f)(\bar{\eta}(\chi(\sigma_E)(x))) = \bar{\eta}(H)(\bar{\eta}(x), \bar{\eta}(f(x)))]$$

$$= [\bar{\eta}(f)(\bar{\eta}[\chi(\sigma_E)](x)) = \bar{\eta}(H)(x, \bar{\eta}(f)(x))]$$

$$= [\bar{\eta}(f)(\sigma_E(x)) = \bar{\eta}(H)(x, \bar{\eta}(f)(x))],$$

and, replacing f by g in the case of $\bar{\eta}(\mathfrak{w}_{\ell})$, we obtain

$$\bar{\eta}(\mathfrak{w}_{m}) = [\bar{\eta}(g)(\sigma_E(x)) = \bar{\eta}(H)(x, \bar{\eta}(g)(x))],$$

$$\bar{\eta}(\mathfrak{w}_{n+1}) = \bar{\eta}(f(x) = g(x))$$

$$= [\bar{\eta}(f(x)) = \bar{\eta}(g(x))]$$

$$= [\bar{\eta}(f)(x) = \bar{\eta}(g)(x)],$$

and (see part (XI) of §4.4) we see that $\bar{\eta}(\mathfrak{w}_{n+1})$ can be obtained from $\bar{\eta}(\mathfrak{w}_{k})$, $\bar{\eta}(\mathfrak{w}_{\ell})$, and $\bar{\eta}(\mathfrak{w}_{m})$ by inference rule (IND) of $\mathrm{EAR}_E^{\mathrm{rec}}$.

Fourth, assume that for some k, $1 \leq k \leq n$, the wff $\mathfrak{w}_{n+1}$ is obtained from $\mathfrak{w}_k$ by inference rule (S_1) (applied to wffs of $\mathrm{EAR}_{\eta SE}^{\chi\text{-rec}}$). Then for functions $f(x)$ and $g(x)$ of $\mathrm{EAR}_{\eta SE}^{\chi\text{-rec}}$ (where each of $f(x)$ and $g(x)$ may have parameters in addition to x, where both functions have the same parameters) and term T of $\mathrm{EAR}_{\eta SE}^{\chi\text{-rec}}$ we have

$$\mathfrak{w}_k = [f(x) = g(x)], \qquad \mathfrak{w}_{n+1} = [f(T) = g(T)].$$

Then by Lemma 2 (which is in §4.19) and Proposition 1α of §4.15 we have

$$\bar{\eta}(\mathfrak{w}_k) = \bar{\eta}(f(x) = g(x))$$
$$= [\bar{\eta}(f(x)) = \bar{\eta}(g(x))]$$
$$= [\bar{\eta}(f)(x) = \bar{\eta}(g)(x)],$$

$$\bar{\eta}(\mathfrak{w}_{n+1}) = \bar{\eta}(f(T) = g(T))$$
$$= [\bar{\eta}(f(T)) = \bar{\eta}(g(T))]$$
$$= [\bar{\eta}(f)(\bar{\eta}(T)) = \bar{\eta}(g)(\bar{\eta}(T))],$$

and (see part (XI) of §4.4) we see that $\bar{\eta}(\mathfrak{w}_{n+1})$ can be obtained from $\bar{\eta}(\mathfrak{w}_k)$ by inference rule (S_1) of EAR_E^{rec}.

<u>Fifth</u>, assume that for some k, $1 \le k \le n$, the wff $\mathfrak{w}_{n+1}$ is obtained from $\mathfrak{w}_k$ by inference rule (S_2) (applied to wffs of $EAR_{\eta SE}^{\chi\text{-rec}}$). Then for terms T and U and function $f(x)$ of $EAR_{\eta SE}^{\chi\text{-rec}}$ (where the function f may have parameters in addition to x), we have

$$\mathfrak{w}_k = [T = U], \qquad \mathfrak{w}_{n+1} = [f(T) = f(U)].$$

Then by Lemma 2 (which is in §4.19) and Proposition 1α of §4.15 we have

$$\bar{\eta}(\mathfrak{w}_k) = \bar{\eta}(T = U) = [\bar{\eta}(T) = \bar{\eta}(U)],$$
$$\bar{\eta}(\mathfrak{w}_{n+1}) = \bar{\eta}(f(T) = f(U))$$
$$= [\bar{\eta}(f(T)) = \bar{\eta}(f(U))]$$
$$= [\bar{\eta}(f)(\bar{\eta}(T)) = \bar{\eta}(f)(\bar{\eta}(U))],$$

and (see part (XI) of §4.4) we see that $\bar{\eta}(\mathfrak{w}_{n+1})$ can be obtained from $\bar{\eta}(\mathfrak{w}_k)$ by inference rule (S_2) of EAR_E^{rec}.

Sixth, assume that $\mathfrak{w}_{n+1}$ is obtained from certain of the wffs $\mathfrak{w}_1, \mathfrak{w}_2, \ldots, \mathfrak{w}_n$ by inference rule (U) of $\mathrm{EAR}_{\eta SE}^{\chi\text{-rec}}$ [1]. In turn (see footnote 2 to the Uniqueness Rule (U) in item (4) of part (XI) of §4.4), for some $m \geq 2$, let $\mathfrak{v}_1, \mathfrak{v}_2, \ldots, \mathfrak{v}_m$ be the conditional equations associated with the derivation of $\mathfrak{w}_{n+1}$.

By the Uniqueness Rule (U) it follows that we are assuming that:

(a) $\mathfrak{w}_n = (\#_S|x)\mathfrak{w}_{n+1}$ [2];

(b) $\mathfrak{v}_1 = \mathfrak{w}_{n+1}$ and, further, $\mathfrak{v}_1$ is assumed as a hypothesis;

(c) for $i = 2, 3, \ldots, m$, the wff $\mathfrak{v}_i$ is derived from one of $\mathfrak{w}_1, \mathfrak{w}_2, \ldots, \mathfrak{w}_n, \mathfrak{v}_1, \mathfrak{v}_2, \ldots, \mathfrak{v}_{i-1}$ by either inference rule (S_1) or inference rule (S_2) (cf. item (3) of part (XI) of §4.4);

(d) $\mathfrak{v}_m = (\chi(\sigma_E)(x)|x)\mathfrak{w}_{n+1}$ (cf. Uniqueness Rule (U)).

Recalling relations (5) and (6) in the proof of Proposition 1 of §4.19, it follows from condition (a) that

$$\overline{\eta}(\mathfrak{w}_n) = \overline{\eta}((\#_S|x)\mathfrak{w}_{n+1}) = (\overline{\eta}(\#_S)|x)\overline{\eta}(\mathfrak{w}_{n+1}) = (0|x)\overline{\eta}(\mathfrak{w}_{n+1})$$

---

(1) Recall that, as mentioned above, the inference rules of $\mathrm{EAR}_{\eta SE}^{\chi\text{-rec}}$ are the inference rules of $\mathrm{EAR}_{E}^{\text{rec}}$ (applied to wffs of $\mathrm{EAR}_{\eta SE}^{\chi\text{-rec}}$).

(2) For convenience, in order to avoid the need for using a metatheoretical variable, we assume that x is a numerical variable occurring in $\mathfrak{w}_{n+1}$.

(we have $\eta(\#_S) = 0$ by Definition 1 of §4.15), and it follows from condition (d) that

$$\begin{aligned}\bar{\eta}(\mathfrak{v}_m) &= \bar{\eta}((\chi(\sigma_E)(x)\,|x)\mathfrak{w}_{n+1}) \\ &= (\bar{\eta}(\chi(\sigma_E)(x))\,|x)\bar{\eta}(\mathfrak{w}_{n+1}) \\ &= (\eta[\chi(\sigma_E)](x)\,|x)\bar{\eta}(\mathfrak{w}_{n+1}) \\ &= (\sigma_E(x)\,|x)\bar{\eta}(\mathfrak{w}_{n+1})\end{aligned}$$

(we have $\eta[\chi(\sigma_E)] = \sigma_E$ by Definition 5 of §4.15).

By condition (c) it follows that for each i, $2 \le i \le m$, the wff $\mathfrak{v}_i$ is obtained from one of $\mathfrak{w}_1$, $\mathfrak{w}_2$, ..., $\mathfrak{w}_n$, $\mathfrak{v}_1$, $\mathfrak{v}_2$, ..., $\mathfrak{v}_{i-1}$ by either inference rule (S_1) or inference rule (S_2). If $\mathfrak{x}$ denotes whichever of $\mathfrak{w}_1$, $\mathfrak{w}_2$, ..., $\mathfrak{w}_n$, $\mathfrak{v}_1$, $\mathfrak{v}_2$, ..., $\mathfrak{v}_{i-1}$ the wff $\mathfrak{v}_i$ is obtained from by one of inference rules (S_1), (S_2) (applied to wffs of $EAR^{\chi\text{-rec}}_{\eta SE}$), then using the reasoning used in the items labeled "Fourth" and "Fifth" here in the equational case we see that we can obtain $\bar{\eta}(\mathfrak{v}_i)$ from $\bar{\eta}(\mathfrak{x})$ by either inference rule (S_1) or inference rule (S_2) (of EAR^{rec}_E).

We thus see that $\bar{\eta}(\mathfrak{w}_{n+1})$ can be obtained from certain of the wffs $\bar{\eta}(\mathfrak{w}_1)$, $\bar{\eta}(\mathfrak{w}_2)$, ..., $\bar{\eta}(\mathfrak{w}_n)$ by the Uniqueness Rule (U) of EAR^{rec}_E (see §4.4), with conditional equations $\bar{\eta}(\mathfrak{v}_1)$, $\bar{\eta}(\mathfrak{v}_2)$, ..., $\bar{\eta}(\mathfrak{v}_m)$.

Thus in any event we see that in the equational case the wff $\bar{\eta}(\mathfrak{w}_{n+1})$ can be obtained from certain of the wffs $\bar{\eta}(\mathfrak{w}_1)$, $\bar{\eta}(\mathfrak{w}_2)$, ..., $\bar{\eta}(\mathfrak{w}_n)$ by an inference rule of EAR^{rec}_E.

This completes the proof of the theorem.

$\mathscr{F}_{BS}$-THEOREM 2. <u>Let S be an atomic infinite-dimensional non-zeroed semiological space, let ηSE be a Peano semicarrier space, let $\bar{\eta}$ be the canonical recursive-extension of η to wffs, and let $\chi: \mathbf{F}_E^{rec} \longrightarrow \mathbf{F}_S^{total}$ be a total function of the type specified in</u> Definition 5 of §4.15. <u>Then, for each $n \geq 1$, if</u>

$$\mathfrak{u}_1, \mathfrak{u}_2, \ldots, \mathfrak{u}_n$$

<u>is a proof in AR_E^{rec}, then there exists a proof</u>

$$\mathfrak{w}_1, \mathfrak{w}_2, \ldots, \mathfrak{w}_n$$

<u>in $\mathrm{AR}_{\eta SE}^{\chi\text{-rec}}$ such that</u> $\bar{\eta}(\mathfrak{w}_i) = \mathfrak{u}_i$ $(1 \leq i \leq n)$.

<u>Proof</u>. We prove the theorem using induction on n. First, if $\mathfrak{u}_1$ is a proof in AR_E^{rec}, then $\mathfrak{u}_1$ is either an explicit axiom or an instance of an axiom schema of AR_E^{rec}. It then follows by Lemmas 1, 3, and 4 (which are in §4.19) that there exists a wff $\mathfrak{w}_1$ of $\mathrm{AR}_{\eta SE}^{\chi\text{-rec}}$ such that $\bar{\eta}(\mathfrak{w}_1) = \mathfrak{u}_1$ and such that $\mathfrak{w}_1$ is an explicit axiom of $\mathrm{AR}_{\eta SE}^{\chi\text{-rec}}$ or an instance of an axiom schema of $\mathrm{AR}_{\eta SE}^{\chi\text{-rec}}$, in which case $\mathfrak{w}_1$ is a proof in $\mathrm{AR}_{\eta SE}^{\chi\text{-rec}}$.

For some $n \geq 1$ assume as induction hypothesis that the theorem holds for proofs in AR_E^{rec} of length n and assume that $\mathfrak{u}_1, \mathfrak{u}_2, \ldots, \mathfrak{u}_n, \mathfrak{u}_{n+1}$ is a proof in AR_E^{rec}. Then $\mathfrak{u}_1, \mathfrak{u}_2, \ldots, \mathfrak{u}_n$ is also a proof in AR_E^{rec}, and so by the induction hypothesis there exists a proof $\mathfrak{w}_1, \mathfrak{w}_2, \ldots, \mathfrak{w}_n$ in $\mathrm{AR}_{\eta SE}^{\chi\text{-rec}}$ such that $\bar{\eta}(\mathfrak{w}_i) = \mathfrak{u}_i$ $(i = 1, 2, \ldots, n)$. By Proposition 19 of §4.15 it follows that there exists a unique wff $\mathfrak{w}_{n+1}$ of $\mathrm{AR}_{\eta SE}^{\chi\text{-rec}}$ such that $\bar{\eta}(\mathfrak{w}_{n+1}) = \mathfrak{u}_{n+1}$.

First, if $\mathfrak{u}_{n+1}$ is either an explicit axiom or instance of an axiom schema of AR_E^{rec}, then by Lemmas 1, 3, and 4 (which are in §4.19) it follows that there exists a wff $\mathfrak{w}_{n+1}$ of $AR_{\eta SE}^{\chi\text{-rec}}$ such that $\bar{\eta}(\mathfrak{w}_{n+1}) = \mathfrak{u}_{n+1}$ and such that $\mathfrak{w}_{n+1}$ is either an explicit axiom of $AR_{\eta SE}^{\chi\text{-rec}}$ or an instance of an axiom schema of $AR_{\eta SE}^{\chi\text{-rec}}$. We see that $\bar{\eta}(\mathfrak{w}_1)$, $\bar{\eta}(\mathfrak{w}_2)$, ..., $\bar{\eta}(\mathfrak{w}_n)$, $\bar{\eta}(\mathfrak{w}_{n+1})$ is trivially a proof in $AR_{\eta SE}^{\chi\text{-rec}}$.

Alternatively, assume that $\mathfrak{u}_{n+1}$ is obtained from certain of the wffs $\mathfrak{u}_1$, $\mathfrak{u}_2$, ..., $\mathfrak{u}_n$ using an inference rule of AR_E^{rec}. We have two cases to consider:

The Sentential Case. Assume that we are dealing with sentential arithmetics, in which case instead of AR we may write SAR. Then by part (VI) of Definition 9 of §4.15 we see that the inference rules of $SAR_{\eta SE}^{\chi\text{-rec}}$ consist of the inference rules of SAR_E^{rec} (applied to wffs of $SAR_{\eta SE}^{\chi\text{-rec}}$), where the successor function σ_E is replaced by the function $\chi(\sigma_E)$, which (see part (XI) of §4.2) implies that there are three inference rules that we have to consider:

<u>First</u>, assume that for some k, $1 \le k \le n$, and some ℓ, $1 \le \ell \le n$, the wff $\mathfrak{u}_{n+1}$ is obtained from $\mathfrak{u}_k$ and $\mathfrak{u}_\ell$ by modus ponens, in which case we may take $\mathfrak{u}_k$ to be the wff $\mathfrak{u}_\ell \Rightarrow \mathfrak{u}_{n+1}$. We let $\mathfrak{w}_{n+1}$ denote the unique wff in $\mathrm{Wff}(SAR_{\eta SE}^{\chi\text{-rec}})$ that is mapped by $\bar{\eta}$ to $\mathfrak{u}_{n+1}$ (such a wff exists by Proposition 19 of §4.15). By the induction hypothesis we have $\bar{\eta}(\mathfrak{w}_k) = \mathfrak{u}_k$ and $\bar{\eta}(\mathfrak{w}_\ell) = \mathfrak{u}_\ell$. From the preceding we have

$$\bar{\eta}(\mathfrak{w}_k) = \mathfrak{u}_k = [\mathfrak{u}_\ell \Rightarrow \mathfrak{u}_{n+1}] = [\bar{\eta}(\mathfrak{w}_\ell) \Rightarrow \bar{\eta}(\mathfrak{w}_{n+1})],$$

and by Proposition 1α of §4.15 we have

$$\bar{\eta}(\mathfrak{w}_\ell \Rightarrow \mathfrak{w}_{n+1}) = [\bar{\eta}(\mathfrak{w}_\ell) \Rightarrow \bar{\eta}(\mathfrak{w}_{n+1})],$$

and from these two results we have $\bar{\eta}(\mathfrak{w}_k) = \bar{\eta}(\mathfrak{w}_\ell \Rightarrow \mathfrak{w}_{n+1}) = \mathfrak{u}_k$. By the uniqueness condition of Proposition 19 of §4.15 (mentioned above) we see that there is a <u>unique</u> wff of $\mathrm{SAR}_{\eta SE}^{\chi\text{-rec}}$ that is mapped by $\bar{\eta}$ to $\mathfrak{u}_k$, and so we have

$$\mathfrak{w}_k = [\mathfrak{w}_\ell \Rightarrow \mathfrak{w}_{n+1}],$$

so that $\mathfrak{w}_{n+1}$ can be obtained from $\mathfrak{w}_k$ and $\mathfrak{w}_\ell$ by modus ponens, which is an inference rule of $\mathrm{SAR}_{\eta SE}^{\chi\text{-rec}}$.

<u>Second</u>, assume that for some k, $1 \le k \le n$, the wff $\mathfrak{u}_{n+1}$ is obtained from $\mathfrak{u}_k$ by substitution of a term T_0 of $\mathrm{SAR}_E^{\mathrm{rec}}$ for a variable x. So we can let $\mathfrak{u}_k$ be a relation $R_0(x)$ of $\mathrm{SAR}_E^{\mathrm{rec}}$ and we have

$$\mathfrak{u}_{n+1} = (T_0|x)\mathfrak{u}_k = R_0(T_0).$$

By part (iv) of Lemma 2 (which is in §4.19) we see that there exist a term T and a relation $R(x)$ of $\mathrm{AR}_{\eta SE}^{\chi\text{-rec}}$ such that $\bar{\eta}(R) = R_0$ and $\bar{\eta}(T) = T_0$ and such that

$$\bar{\eta}(R(T)) = \bar{\eta}(R)(\bar{\eta}(T)) = R_0(T_0) = \mathfrak{u}_{n+1}.$$

So we set $\mathfrak{w}_{n+1} = R(T)$.

By part (iv) of Lemma 2 (which is in §4.19) and relations (5) and (6) in the proof of Proposition 1 we have

$$\bar{\eta}(R(x)) = \bar{\eta}(R)(\bar{\eta}(x)) = \bar{\eta}(R)(\eta(x)) = \bar{\eta}(R)(x) = R_0(x) = \mathfrak{u}_k$$

(we have $\eta(x) = x$ by relation (5) in the proof of Proposition 1 in §4.19, and with $\overline{\eta}(\mathfrak{w}_k) = \mathfrak{u}_k$ it follows by the uniqueness condition of Proposition 19 of §4.15 that $\mathfrak{w}_k = R(x)$. Thus

$$(T\,|x)\mathfrak{w}_k = (T\,|x)R(x) = R(T) = \mathfrak{w}_{n+1},$$

and so $\mathfrak{w}_{n+1}$ can be obtained from $\mathfrak{w}_k$ by substitution of a term of $\mathrm{SAR}^{\chi\text{-rec}}_{\eta SE}$ for a numerical variable.

<u>Third</u>, assume that for some k, $1 \le k \le n$, and some ℓ, $1 \le \ell \le n$, the wff $\mathfrak{u}_{n+1}$ is obtained from $\mathfrak{u}_k$ and $\mathfrak{u}_\ell$ by induction. Then (see item (2) in part (XI) of §4.2) $\mathfrak{u}_{n+1}$ is some relation $R_0(x)$ of $\mathrm{SAR}^{\mathrm{rec}}_E$ and $\mathfrak{u}_k$ and $\mathfrak{u}_\ell$ are the relations $R_0(0)$ and

$$R_0(x) \rightarrow R_0(\sigma_E(x)),$$

respectively. By part (iv) of Lemma 2 (which is in §4.19) we see that there exists a relation $R(x)$ of $\mathrm{SAR}^{\chi\text{-rec}}_{\eta SE}$ such that $\overline{\eta}(R) = R_0$, so that

$$\overline{\eta}(R(x)) = \overline{\eta}(R)(\overline{\eta}(x)) = \overline{\eta}(R)(\eta(x)) = R_0(x) = \mathfrak{u}_{n+1}.$$

So we set $\mathfrak{w}_{n+1} = R(x)$.

By part (iii) of Lemma 2 (which is in §4.19) we have

$$\overline{\eta}(R(\#_S)) = \overline{\eta}(R)(\overline{\eta}(\#_S)) = \overline{\eta}(R)(0) = R_0(0) = \mathfrak{u}_k$$

(we have $\overline{\eta}(\#_S) = \eta(\#_S) = \#_E = 0$ by relation (5) of the proof of Proposition 1 of §4.19 and by Definition 1 of §4.15) and also (by the same reference, along with several other references that we have previously used in this proof)

$$\bar{\eta}(R(x) \Rightarrow R(\chi(\sigma_E)(x))) = [\bar{\eta}(R(x)) \Rightarrow \bar{\eta}(R(\chi(\sigma_E)(x))))]$$

$$= [\bar{\eta}(R)(x) \Rightarrow \bar{\eta}(R)(\bar{\eta}(\chi(\sigma_E)(x)))]$$

$$= [R_0(x) \rightarrow R_0(\eta[\chi(\sigma_E)](x))]$$

$$= [R_0(x) \rightarrow R_0(\sigma_E(x))]$$

$$= \mathfrak{u}_\ell.$$

We also have $\bar{\eta}(\mathfrak{w}_k) = \mathfrak{u}_k$ and $\bar{\eta}(\mathfrak{w}_\ell) = \mathfrak{u}_\ell$ (from the induction hypothesis), and so from the uniqueness condition of Proposition 19 of §4.15 we see that $\mathfrak{w}_k = R(\#_S)$ and

$$\mathfrak{w}_\ell = [R(x) \rightarrow R(\chi(\sigma_E)(x))].$$

Thus we see that the wff $\mathfrak{w}_{n+1}$ (= $R(x)$) can be derived from wffs $\mathfrak{w}_k$ and $\mathfrak{w}_\ell$ by induction, which is an inference rule of $\mathrm{SAR}^{\chi\text{-rec}}_{\eta SE}$.

Thus in any event we see that in the sentential case the wff $\mathfrak{w}_{n+1}$ can be derived from certain of the wffs $\mathfrak{w}_1$, $\mathfrak{w}_2$, ..., $\mathfrak{w}_n$ by the inference rules of $\mathrm{SAR}^{\chi\text{-rec}}_{\eta SE}$.

The Equational Case. Assume that we are dealing with equational arithmetics, in which case instead of AR we may write EAR. Then by part (VI) of Definition 9 of §4.15 we see that the inference rules of $\mathrm{EAR}^{\chi\text{-rec}}_{\eta SE}$ consist of the inference rules of $\mathrm{EAR}^{\mathrm{rec}}_{E}$ (applied to wffs of $\mathrm{EAR}^{\chi\text{-rec}}_{\eta SE}$), where the successor function σ_E is replaced by the function $\chi(\sigma_E)$, which (see part (XI) of §4.4) implies that there are six inference rules to consider:

<u>First</u>, assume that for some k, $1 \le k \le n$, the wff $\mathfrak{u}_{n+1}$ is obtained from $\mathfrak{u}_k$ by inference rule (E_1). Then for terms $\mathbf{t}'_1$ and $\mathbf{t}'_2$ of EAR_E^{rec} we have

$$\mathfrak{u}_k = [\mathbf{t}'_1 = \mathbf{t}'_2], \qquad \mathfrak{u}_{n+1} = [\mathbf{t}'_2 = \mathbf{t}'_1].$$

By Proposition 19 of §4.15 there exist unique

$$\mathfrak{w}_k,\ \mathfrak{w}_{n+1} \in Wff(EAR_{\eta SE}^{\chi\text{-}rec})$$

such that $\bar{\eta}(\mathfrak{w}_k) = \mathfrak{u}_k$ and $\bar{\eta}(\mathfrak{w}_{n+1}) = \mathfrak{u}_{n+1}$.

By part (ii) of Lemma 2 (which is in §4.19) there exist terms $\mathbf{t}_1$ and $\mathbf{t}_2$ of $EAR_{\eta SE}^{\chi\text{-}rec}$ such that $\bar{\eta}(\mathbf{t}_1) = \mathbf{t}'_1$ and $\bar{\eta}(\mathbf{t}_2) = \mathbf{t}'_2$. Thus $\mathbf{t}_1 = \mathbf{t}_2$ and $\mathbf{t}_2 = \mathbf{t}_1$ are wffs of $EAR_{\eta SE}^{\chi\text{-}rec}$ and by Proposition 1α of §4.15 we have

$$\bar{\eta}(\mathbf{t}_1 = \mathbf{t}_2) = [\bar{\eta}(\mathbf{t}_1) = \bar{\eta}(\mathbf{t}_2)] = [\mathbf{t}'_1 = \mathbf{t}'_2] = \mathfrak{u}_k = \bar{\eta}(\mathfrak{w}_k),$$

$$\bar{\eta}(\mathbf{t}_2 = \mathbf{t}_1) = [\bar{\eta}(\mathbf{t}_2) = \bar{\eta}(\mathbf{t}_1)] = [\mathbf{t}'_2 = \mathbf{t}'_1] = \mathfrak{u}_{n+1} = \bar{\eta}(\mathfrak{w}_{n+1}).$$

Then by the uniqueness condition of Proposition 19 of §4.15 we obtain $\mathfrak{w}_k = [\mathbf{t}_1 = \mathbf{t}_2]$ and $\mathfrak{w}_{n+1} = [\mathbf{t}_2 = \mathbf{t}_1]$, and we see (cf. part (XI) of §4.4) that $\mathfrak{w}_{n+1}$ can be obtained from $\mathfrak{w}_k$ by inference rule (E_1), which (when applied to wffs of $EAR_{\eta SE}^{\chi\text{-}rec}$) is a wff of that arithmetic.

<u>Second</u>, assume that for some k, $1 \le k \le n$, and some ℓ, $1 \le \ell \le n$, the wff $\mathfrak{u}_{n+1}$ is obtained from $\mathfrak{u}_k$ and $\mathfrak{u}_\ell$ by inference rule (E_2). Then for terms $\mathbf{t}'_1$, $\mathbf{t}'_2$, and $\mathbf{t}'_3$ of EAR_E^{rec} we have

$$\mathfrak{u}_k = [\mathbf{t}'_1 = \mathbf{t}'_2], \qquad \mathfrak{u}_\ell = [\mathbf{t}'_2 = \mathbf{t}'_3], \qquad \mathfrak{u}_{n+1} = [\mathbf{t}'_1 = \mathbf{t}'_3].$$

By Proposition 19 of §4.15 there exist unique wffs

$$\mathfrak{w}_k,\ \mathfrak{w}_\ell,\ \mathfrak{w}_{n+1} \in \mathrm{Wff}(\mathrm{EAR}^{\chi\text{-rec}}_{\eta SE})$$

such that $\bar{\eta}(\mathfrak{w}_k) = \mathfrak{u}_k$, $\bar{\eta}(\mathfrak{w}_\ell) = \mathfrak{u}_\ell$, and $\bar{\eta}(\mathfrak{w}_{n+1}) = \mathfrak{u}_{n+1}$.

By part (ii) of Lemma 2 (which is in §4.19) there exist terms $\mathbf{t}_1$, $\mathbf{t}_2$, and $\mathbf{t}_3$ of $\mathrm{EAR}^{\chi\text{-rec}}_{\eta SE}$ such that $\bar{\eta}(\mathbf{t}_i) = \mathbf{t}'_i$ ($i = 1, 2, 3$). Thus $\mathbf{t}_1 = \mathbf{t}_2$, $\mathbf{t}_2 = \mathbf{t}_3$, and $\mathbf{t}_1 = \mathbf{t}_3$ are wffs of $\mathrm{EAR}^{\chi\text{-rec}}_{\eta SE}$, and by Proposition 1α of §4.15 we have

$$\bar{\eta}(\mathbf{t}_1 = \mathbf{t}_2) = [\bar{\eta}(\mathbf{t}_1) = \bar{\eta}(\mathbf{t}_2)] = [\mathbf{t}'_1 = \mathbf{t}'_2] = \mathfrak{u}_k = \bar{\eta}(\mathfrak{w}_k),$$

$$\bar{\eta}(\mathbf{t}_2 = \mathbf{t}_3) = [\bar{\eta}(\mathbf{t}_2) = \bar{\eta}(\mathbf{t}_3)] = [\mathbf{t}'_2 = \mathbf{t}'_3] = \mathfrak{u}_\ell = \bar{\eta}(\mathfrak{w}_\ell),$$

$$\bar{\eta}(\mathbf{t}_1 = \mathbf{t}_3) = [\bar{\eta}(\mathbf{t}_1) = \bar{\eta}(\mathbf{t}_3)] = [\mathbf{t}'_1 = \mathbf{t}'_3] = \mathfrak{u}_{n+1} = \bar{\eta}(\mathfrak{w}_{n+1}).$$

Then by the uniqueness condition of Proposition 19 of §4.15 mentioned above we obtain $\mathfrak{w}_k = [\mathbf{t}_1 = \mathbf{t}_2]$, $\mathfrak{w}_\ell = [\mathbf{t}_2 = \mathbf{t}_3]$, and $\mathfrak{w}_{n+1} = [\mathbf{t}_1 = \mathbf{t}_3]$, and we see (cf. part (XI) of §4.4) that $\mathfrak{w}_{n+1}$ can be obtained from $\mathfrak{w}_k$ and $\mathfrak{w}_\ell$ by inference rule (E_2) of $\mathrm{EAR}^{\chi\text{-rec}}_{\eta SE}$ (applied to wffs of $\mathrm{EAR}^{\chi\text{-rec}}_{\eta SE}$).

<u>Third</u>, assume that for some k, $1 \le k \le n$, some ℓ, $1 \le \ell \le n$, and some m, $1 \le m \le n$, the wff $\mathfrak{u}_{n+1}$ is obtained from $\mathfrak{u}_k$, $\mathfrak{u}_\ell$, and $\mathfrak{u}_m$ by inference rule (IND). Then for functions $f_0(x)$, $g_0(x)$, and $H_0(x,y)$ of $\mathrm{EAR}^{\mathrm{rec}}_E$ (where each of f_0, g_0, and H_0 may have parameters in addition to x and y, with each of f_0, g_0, and H_0 having exactly the same parameters) we have

$$\mathfrak{u}_k = [f_0(0) = g_0(0)], \qquad \mathfrak{u}_\ell = [f_0(\sigma_E(x)) = H_0(x, f_0(x))],$$

$$\mathfrak{u}_{n+1} = [f_0(x) = g_0(x)], \qquad \mathfrak{u}_m = [g_0(\sigma_E(x)) = H_0(x, g_0(x))],$$

By Proposition 19 of §4.15 there exist unique

$$\mathfrak{w}_k,\ \mathfrak{w}_\ell,\ \mathfrak{w}_m,\ \mathfrak{w}_{n+1} \in \mathrm{Wff}(\mathrm{EAR}^{\chi\text{-rec}}_{\eta SE})$$

such that $\overline{\eta}(\mathfrak{w}_k) = \mathfrak{u}_k$, $\overline{\eta}(\mathfrak{w}_\ell) = \mathfrak{u}_\ell$, $\overline{\eta}(\mathfrak{w}_m) = \mathfrak{u}_m$, and $\overline{\eta}(\mathfrak{w}_{n+1}) = \mathfrak{u}_{n+1}$.

By part (iv) of Lemma 2 (which is in §4.19) there exist functions f, g, and H of $\mathrm{EAR}^{\chi\text{-rec}}_{\eta SE}$ (with exactly the same parameters as f_0, g_0, and H_0, respectively) such that $\overline{\eta}(f) = f_0$, $\overline{\eta}(g) = g_0$, and $\overline{\eta}(H) = H_0$. We see that

$$f(\#_S) = g(\#_S),$$

$$f(\chi(\sigma_E)(x)) = H(x, f(x)),$$

$$g(\chi(\sigma_E)(x)) = H(x, g(x)),$$

$$f(x) = g(x)$$

are wffs of $\mathrm{EAR}^{\chi\text{-rec}}_{\eta SE}$. So, using reasoning used previously in this proof, we obtain

$$\begin{aligned}\overline{\eta}(f(\#_S) = g(\#_S)) &= [\overline{\eta}(f(\#_S)) = \overline{\eta}(g(\#_S))]\\ &= [\overline{\eta}(f)(\eta(\#_S)) = \overline{\eta}(g)(\eta(\#_S))]\\ &= [f_0(0) = g_0(0)]\\ &= \mathfrak{u}_k\\ &= \overline{\eta}(\mathfrak{w}_k),\end{aligned}$$

$$\begin{aligned}\overline{\eta}(f(\chi(\sigma_E)(x)) = H(x, f(x))) &\\ = [\overline{\eta}(f(\chi(\sigma_E)(x))) &= \overline{\eta}(H(x, f(x)))]\end{aligned}$$

$$= [\bar{\eta}(f)(\bar{\eta}(\chi(\sigma_E)(x))) = \bar{\eta}(H)(\bar{\eta}(x),\bar{\eta}(f(x)))]$$

$$= [f_0(\eta[\chi(\sigma_E)](\bar{\eta}(x))) = H_0(x,\bar{\eta}(f)(x))]$$

$$= [f_0(\sigma_E(x)) = H_0(x,f(x))]$$

$$= \mathfrak{u}_\ell$$

$$= \bar{\eta}(\mathfrak{w}_\ell),$$

and, replacing f and f_0 by g and g_0, respectively, in the preceding item, we obtain

$$\bar{\eta}(g(\chi(\sigma_E)(x)) = H(x,g(x)))$$

$$= [\bar{\eta}(g)(\sigma_E(x)) = H_0(x,g_0(x))]$$

$$= \mathfrak{u}_m$$

$$= \bar{\eta}(\mathfrak{w}_m),$$

and, lastly, we have

$$\bar{\eta}(f(x) = g(x)) = [\bar{\eta}(f(x)) = \bar{\eta}(g(x))]$$

$$= [\bar{\eta}(f)(x) = \bar{\eta}(g)(x)]$$

$$= [f_0(x) = g_0(x)]$$

$$= \mathfrak{u}_{n+1}$$

$$= \bar{\eta}(\mathfrak{w}_{n+1}).$$

From the preceding, with the uniqueness condition of Proposition 19 of §4.15 mentioned above, we have

$$\mathfrak{w}_k = [f(\#_S) = g(\#_S)],$$

$$\mathfrak{w}_\ell = [f(\chi(\sigma_E)(x)) = H(x,f(x))],$$

$$\mathfrak{w}_m = [g(\chi(\sigma_E)(x)) = H(x, g(x))],$$

$$\mathfrak{w}_{n+1} = [f(x) = g(x)],$$

and so $\mathfrak{w}_{n+1}$ can be obtained from $\mathfrak{w}_k$, $\mathfrak{w}_\ell$, and $\mathfrak{w}_m$ by inference rule (IND) of $\text{EAR}^{\chi\text{-rec}}_{\eta SE}$.

Fourth, assume that for some k, $1 \le k \le n$, the wff $\mathfrak{u}_{n+1}$ is obtained from $\mathfrak{u}_k$ by inference rule (S_1) of $\text{EAR}^{\text{rec}}_E$. Then for functions $f_0(x)$ and $g_0(x)$ of $\text{EAR}^{\text{rec}}_E$ (where each of $f_0(x)$ and $g_0(x)$ may have parameters in addition to x, where both functions have the same parameters) and for term T_0 of $\text{EAR}^{\text{rec}}_E$ we have

$$\mathfrak{u}_k = [f_0(x) = g_0(x)], \qquad \mathfrak{u}_{n+1} = [f_0(T_0) = g_0(T_0)].$$

By Proposition 19 of §4.15 there exist unique

$$\mathfrak{w}_k,\ \mathfrak{w}_{n+1} \in \text{Wff}(\text{EAR}^{\chi\text{-rec}}_{\eta SE})$$

such that $\overline{\eta}(\mathfrak{w}_k) = \mathfrak{u}_k$ and $\overline{\eta}(\mathfrak{w}_{n+1}) = \mathfrak{u}_{n+1}$.

By part (iv) of Lemma 2 (which is in §4.19) there exist functions $f(x)$ and $g(x)$ of $\text{EAR}^{\chi\text{-rec}}_{\eta SE}$ and a term T of $\text{EAR}^{\chi\text{-rec}}_{\eta SE}$ such that $\overline{\eta}(f) = f_0$, $\overline{\eta}(g) = g_0$, and $\overline{\eta}(T) = T_0$ and such that

$$\overline{\eta}(f(T)) = \overline{\eta}(f)(\overline{\eta}(T)) \qquad \text{and} \qquad \overline{\eta}(g(T)) = \overline{\eta}(g)(\overline{\eta}(T)).$$

We see that $f(x) = g(x)$ and $f(T) = g(T)$ are wffs of $\text{EAR}^{\chi\text{-rec}}_{\eta SE}$.

Thus, with reasoning previously used in the proof, we have

$$\overline{\eta}(f(x) = g(x)) = [\overline{\eta}(f(x)) = \overline{\eta}(g(x))]$$

$$= [\overline{\eta}(f)(x) = \overline{\eta}(g)(x)]$$

$$= [f_0(x) = g_0(x)]$$

$$= \mathfrak{u}_k$$

$$= \overline{\eta}(\mathfrak{w}_k),$$

$$\overline{\eta}(f(T) = g(T)) = [\overline{\eta}(f(T)) = \overline{\eta}(g(T))]$$

$$= [\overline{\eta}(f)(\overline{\eta}(T)) = \overline{\eta}(g)(\overline{\eta}(T))]$$

$$= [f_0(T_0) = g_0(T_0)]$$

$$= \mathfrak{u}_{n+1}$$

$$= \overline{\eta}(\mathfrak{w}_{n+1}).$$

From the preceding, with the uniqueness condition of Proposition 19 of §4.15 mentioned above, we obtain

$$\mathfrak{w}_k = [f(x) = g(x)], \qquad \mathfrak{w}_{n+1} = [f(T) = g(T)],$$

and so $\mathfrak{w}_{n+1}$ can be obtained from $\mathfrak{w}_k$ by inference rule (S_1) of $EAR^{\chi\text{-rec}}_{\eta SE}$ (cf. part (XI) of §4.4).

Fifth, assume that for some k, $1 \le k \le n$, the wff $\mathfrak{u}_{n+1}$ is obtained from $\mathfrak{u}_k$ by inference rule (S_2) of EAR^{rec}_E. Then for terms T_0 and U_0 of EAR^{rec}_E and function $f_0(x)$ of EAR^{rec}_E (where the function f_0 may have parameters in addition to x) we have

$$\mathfrak{u}_k = [T_0 = U_0], \qquad \mathfrak{u}_{n+1} = [f_0(T_0) = f_0(U_0)].$$

By Proposition 19 of §4.15 there exist unique

$$\mathfrak{w}_k,\ \mathfrak{w}_{n+1} \in \mathrm{Wff}(EAR^{\chi\text{-rec}}_{\eta SE})$$

such that $\overline{\eta}(\mathfrak{w}_k) = \mathfrak{u}_k$ and $\overline{\eta}(\mathfrak{w}_{n+1}) = \mathfrak{u}_{n+1}$.

By part (iv) of Lemma 2 (which is in §4.19) there exist a function $f(x)$ of $\mathrm{EAR}_{\eta SE}^{\chi\text{-rec}}$ and terms T and U of $\mathrm{EAR}_{\eta SE}^{\chi\text{-rec}}$ such that $\overline{\eta}(f) = f_0$, $\overline{\eta}(T) = T_0$, $\overline{\eta}(U) = U_0$,

$$\overline{\eta}(f(T)) = \overline{\eta}(f)(\overline{\eta}(T)), \qquad \overline{\eta}(f(U)) = \overline{\eta}(f)(\overline{\eta}(U)).$$

We see that $T = U$ and $f(T) = f(U)$ are wffs of $\mathrm{EAR}_{\eta SE}^{\chi\text{-rec}}$.

Thus, with reasoning used previously in the proof, we have

$$\overline{\eta}(T = U) = [\overline{\eta}(T) = \overline{\eta}(U)] = [T_0 = U_0] = \mathfrak{u}_k = \overline{\eta}(\mathfrak{w}_k),$$

$$\begin{aligned}\overline{\eta}(f(T) = f(U)) &= [\overline{\eta}(f(T)) = \overline{\eta}(f(U))] \\ &= [\overline{\eta}(f)(\overline{\eta}(T)) = \overline{\eta}(f)(\overline{\eta}(U))] \\ &= [f_0(T_0) = f_0(U_0)] \\ &= \mathfrak{u}_{n+1} \\ &= \overline{\eta}(\mathfrak{w}_{n+1}).\end{aligned}$$

From the preceding, with the uniqueness condition of Proposition 19 of §4.15 mentioned above, we obtain

$$\mathfrak{w}_k = [T = U], \qquad \mathfrak{w}_{n+1} = [f(T) = f(U)],$$

and so $\mathfrak{w}_{n+1}$ can be obtained from $\mathfrak{w}_k$ by inference rule (S_2) of $\mathrm{EAR}_{\eta SE}^{\chi\text{-rec}}$ (cf. part (XI) of §4.4).

<u>Sixth</u>, assume that $\mathfrak{u}_{n+1}$ is obtained from certain of the wffs $\mathfrak{u}_1, \mathfrak{u}_2, \ldots, \mathfrak{u}_n$ by Uniqueness Rule (U) of $\mathrm{EAR}_E^{\mathrm{rec}}$. In turn (see the Uniqueness Rule (U) (and the footnote to it) in item (4) of part (XI) of §4.4), for some $m \geq 2$, let $\mathfrak{v}_1, \mathfrak{v}_2, \ldots, \mathfrak{v}_m$ be the conditional equations associated with the derivation of

$\mathfrak{u}_{n+1}$.

By the Uniqueness Rule (U) it follows that we are assuming that (cf. footnote 2 above in the present section):

(a) $\mathfrak{u}_n = (0|x)\mathfrak{u}_{n+1}$;

(b) $\mathfrak{v}_1 = \mathfrak{u}_{n+1}$ and, further, $\mathfrak{v}_1$ is assumed as a hypothesis;

(c) for $i = 2, 3, \ldots, m$, the wff $\mathfrak{v}_i$ is derived from one of $\mathfrak{u}_1, \mathfrak{u}_2, \ldots, \mathfrak{u}_n, \mathfrak{v}_1, \mathfrak{v}_2, \ldots, \mathfrak{v}_{i-1}$ by either inference rule (S_1) or inference rule (S_2) (cf. item (3) of part (XI) of §4.4);

(d) $\mathfrak{v}_m = (\sigma_E(x)|x)\mathfrak{u}_{n+1}$.

By Proposition 19 of §4.15 there exist unique wffs

$$\mathfrak{w}_1, \mathfrak{w}_2, \ldots, \mathfrak{w}_n, \mathfrak{w}_{n+1}, \mathfrak{y}_1, \mathfrak{y}_2, \ldots, \mathfrak{y}_m \in \mathrm{Wff}(\mathrm{EAR}^{\chi\text{-rec}}_{\eta SE})$$

such that

$$\bar{\eta}(\mathfrak{w}_i) = \mathfrak{u}_i \ (1 \le i \le n+1), \qquad \bar{\eta}(\mathfrak{y}_j) = \mathfrak{v}_j \ (1 \le j \le m).$$

We see that $(\#_S|x)\mathfrak{w}_{n+1}$ and $(\chi(\sigma_E)(x)|x)\mathfrak{w}_{n+1}$ are wffs of $\mathrm{EAR}^{\chi\text{-rec}}_{\eta SE}$. Recalling relations (5) and (6) in the proof of Proposition 1 of §4.19, we see that

$$\bar{\eta}((\#_S|x)\mathfrak{w}_{n+1}) = (\bar{\eta}(\#_S)|x)\bar{\eta}(\mathfrak{w}_{n+1}) = (0|x)\mathfrak{u}_{n+1} = \mathfrak{u}_n = \bar{\eta}(\mathfrak{w}_n),$$

$$\begin{aligned}\bar{\eta}((\chi(\sigma_E)(x)|x)\mathfrak{w}_{n+1}) &= (\bar{\eta}(\chi(\sigma_E)(x))|x)\bar{\eta}(\mathfrak{w}_{n+1}) \\ &= (\eta[\chi(\sigma_E)](x)|x)\mathfrak{u}_{n+1} \\ &= (\sigma_E(x)|x)\mathfrak{u}_{n+1}\end{aligned}$$

$$= \mathfrak{v}_m$$

$$= \bar{\eta}(\mathfrak{y}_m)$$

(we have $\eta(\#_S) = 0$ by Definition 1 of §4.15 and have $\eta[\chi(\sigma_E)] = \sigma_E$ by Definition 5 of §4.15).

From the preceding, with the uniqueness condition of Proposition 19 of §4.15 mentioned above, we obtain

$$\mathfrak{w}_n = (\#_S | x)\mathfrak{w}_{n+1}, \qquad \mathfrak{y}_m = (\chi(\sigma_E)(x) | x)\mathfrak{w}_{n+1}.$$

By condition (b) above we have $\bar{\eta}(\mathfrak{y}_1) = \mathfrak{v}_1 = \mathfrak{u}_{n+1}$ which with $\bar{\eta}(\mathfrak{w}_{n+1}) = \mathfrak{u}_{n+1}$ and the uniqueness condition of Proposition 19 of §4.15 mentioned above yields

$$\mathfrak{y}_1 = \mathfrak{w}_{n+1}.$$

Now let $2 \le i \le m$. Then by condition (c) above it follows that $\mathfrak{v}_i$ is derived from one of $\mathfrak{u}_1, \mathfrak{u}_2, \ldots, \mathfrak{u}_n, \mathfrak{v}_1, \mathfrak{v}_2, \ldots, \mathfrak{v}_{i-1}$ by either inference rule (S_1) or inference rule (S_2). Let $\mathfrak{x}$ denote the wff from which $\mathfrak{v}_i$ is derived. By the uniqueness condition of Proposition 19 of §4.15 mentioned above with respect to the $\mathfrak{w}_j$ and $\mathfrak{v}_k$, there exists a unique wff $\mathfrak{z}$ among $\mathfrak{w}_1, \mathfrak{w}_2, \ldots, \mathfrak{w}_n, \mathfrak{y}_1, \mathfrak{y}_2, \ldots, \mathfrak{y}_{i-1}$ such that $\bar{\eta}(\mathfrak{z}) = \mathfrak{x}$. Then, using the reasoning used in the sections designated "Fourth" and "Fifth" here in the proof of the equational case of the theorem, one shows that the wff $\mathfrak{y}_i$ can be derived from the wff $\mathfrak{z}$ by either inference rule (S_1) or inference rule (S_2).

Thus we have shown that each $\mathfrak{y}_i$ $(2 \le i \le m)$ can be derived from one of $\mathfrak{w}_1, \mathfrak{w}_2, \ldots, \mathfrak{w}_m, \mathfrak{y}_1, \mathfrak{y}_2, \ldots, \mathfrak{y}_{i-1}$ by either inference rule (S_1) or inference rule (S_2) (of $EAR^{\chi\text{-rec}}_{\eta SE}$).

It then follows that $\mathfrak{w}_{\eta+1}$ can be obtained from certain of the wffs $\mathfrak{w}_1, \mathfrak{w}_2, \ldots, \mathfrak{w}_\eta$ by the Uniqueness Rule (U) (cf. item (4) in part (XI) of §4.4) (applied to wffs of $EAR^{\chi\text{-rec}}_{\eta SE}$) with conditional equations $\mathfrak{y}_1, \mathfrak{y}_2, \ldots, \mathfrak{y}_m$.

Thus in any event we see that in the equational case the wff $\mathfrak{w}_{\eta+1}$ can be obtained from certain of the wffs $\mathfrak{w}_1, \mathfrak{w}_2, \ldots,$ $\mathfrak{w}_\eta$ by an inference rule of $EAR^{\chi\text{-rec}}_{\eta SE}$.

This completes the proof.

BIBLIOGRAPHY

Asser, G.

1960. Rekursive Wortfunktionen. Zeitschrift für mathematische Logik und Grundlagen der Mathematik 6, 258-278.

Auslander, L.

1969. What Are Numbers? Scott-Foresman and Co., Glenview, Illinois.

Balzer, W., Moulines, C. U., and Sneed, J. D.

1987. An Architectonic for Science: The Structuralist Program. Reidel, Dordrecht.

Bar-Hillel, Y., et al. (eds.)

1962. Essays on the Foundations of Mathematics. North-Holland, Amsterdam.

Bolzano, B.

1837. Wissenschaftslehre. Seidelchen Buchhandlung, Sulzbach.

Bourbaki, N.

1939-Present. Éléments de Mathématique. Hermann, Paris.

1968. Theory of Sets. Addison-Wesley, Reading, Massachusetts.

1971. Variétés Différentielles et Analytiques: Fascicule de Résultats. Paragraphes 1 à 7, Paragraphes 8 à 15, Hermann, Paris.

1974. Algebra I. Addison-Wesley, Reading, Massachusetts.

Cantor, G.

1895. Beiträge zur Begründung der transfiniten Mengenlehre. Mathematische Annalen 46, 481-512.

Carnap, R.

1962. On the use of Hilbert's ε-operator in scientific theories, in Bar-Hillel [1962], 156-164.

Chevalley, C.

1946. Theory of Lie Groups. Princeton University Press, Princeton, New Jersey.

1956. Fundamental Concepts of Algebra. Academic Press, New York.

Church, A.

1956. Introduction to Mathematical Logic. Princeton University Press, Princeton, New Jersey.

1957. Binary recursive arithmetic. Journal de Mathématiques Pures et Appliquées (9) *36*, 39-55.

Cohen, P. J.

1963-1964. The independence of the continuum hypothesis. Proc. of the NAS **50** (1963), 1143-1148; **51** (1964), 105-110.

Davis, M.

1973. Hilbert's Tenth Problem is Unsolvable. American Mathematical Monthly **80**, 233-269.

Dedekind, R.

1888. Was sind und was sollen die Zahlen? Vieweg, Braunschweig.

1995. What Are Numbers and What Should They Be? (Was sind und was sollen die Zahlen?). Revised, Edited, and Translated by H. Pogorzelski, W. Ryan, and W. Snyder, Research Institute for Mathematics, Orono, Maine.

Dieudonné, J.

1960-1978. Foundations of Modern Analysis: Treatise on Analysis. Academic Press, New York.

1977. Panorama des Mathématique Pures: Le Choix Bourbachique. Gauthier-Villars, Paris.

Fraenkel, A., Bar-Hillel, Y., Levy, A.
1973. Foundations of Set Theory. North-Holland Publishing Co., Amsterdam.

Frege, G.
1884. Die Grundlagen der Arithmetik. W. Koebner, Breslau.
1892. Über Sinn und Bedeutung. Zeitschrift für Philosophie und philosophische Kritik 100, 25-30.
1893-1903. Grundgesetze der Arithmetik. Volumes I, II. Hermann Pohle, Jena.

Gauss, C. F.
1801. Disquisitiones Arithmeticae. Gerhard Fleischner, Leipzig.

Gentzen, G.
1936. Die Widerspruchsfreiheit der reinen Zahlentheorie. Mathematische Annalen 112, 493-565.

Gödel, K.
1940. The Consistency of the Continuum Hypothesis. Princeton University Press, Princeton, New Jersey.
1986. Collected Works. Volume I. Oxford University Press, Oxford.

Goodstein, R. L.
1957. Recursive Number Theory. North-Holland, Amsterdam.

Grothendieck, A., and Dieudonné, J.
1971. Éléments de Géométrie Algébrique. Springer-Verlag, Heidelberg.

Hallett, M.

1984. Cantorian Set Theory and Limitations of Size. Oxford University Press, Oxford.

Hilbert, D., and Bernays, P.

1968-1970. Grundlagen der Mathematik I, II. Zweite Auflage. Springer-Verlag, Heidelberg.

Husserl, E.

1891. Philosophie der Arithmetik. C. E. M. Pfeffer (R. Stricker), Halle.

Kleene, S.

1952. Introduction to Metamathematics. D. Van Nostrand, New York.

Kuratowski, K.

1922. Une méthode d'élimination des nombres transfinis des raisonnements mathématiques. Fundamenta Mathematicae **3**, 76-108.

Landau, E.

1953. Handbuch der Lehre von der Verteilung der Primzahlen. Chelsea Publishing Co., New York.

Lebesgue, H., et al.

1905. Cinq lettres sur la théorie des ensembles. Bulletin de la Société Mathématique de France **33**, 261-273.

Lesniewski, S.

1930. O podstawach matematyki, §§6-9 (On the Foundations of Mathematics, §§6-9). Przeglad Filozoficzny **33**, 77-105.

Mendelson, E.

1973. Number Systems and the Foundations of Analysis. Academic Press, New York.

1979. Introduction to Mathematical Logic. D. Van Nostrand, New York.

Moore, A. W.

1990. The Infinite. Routledge, London.

Mostowski, A.

1939. Über die Unabhängigkeit des Wohlordnungssatzes vom Ordnungsprinzip. Fundamenta Mathematicae **32**, 201-252.

Peano, G.

1889. Arithmetices principia nova methodo exposita. Bocca, Turin.

Pogorzelski, H. A.

1969-1970. Goldbach sentences in abstract arithmetics $\mathcal{A}^k$(A). I, II. Journal für die reine und angewandte Mathematik **237** (1969), 65-96, **243** (1970), 32-54.

1974. On the Goldbach conjecture and the consistency of general recursive arithmetic. Ibidem **268/269**, 1-16.

1976. Dirichlet theorems and prime number hypotheses of a conditional Goldbach theorem. Ibidem **286/287**, 33-45.

1977. Semisemiological structure of the prime numbers and conditional Goldbach theorems. Ibidem **290**, 77-92.

1977a. Goldbach conjecture. Ibidem **292**, 1-12.

Pogorzelski, H. A., and Ryan, W. J.

1982. Foundations of Semiological Theory of Numbers (Foundations of Computability). Volume I, General Semiology I. University of Maine Press, Orono, Maine.

Pogorzelski, H. A., and Ryan, W. J.

1985. Foundations of Semiological Theory of Numbers (Foundations of Computability). Volume II, Semiological Functions. University of Maine Press, Orono, Maine.

Pogorzelski, H. A., and Ryan, W. J. (1)

1988. Foundations of Semiological Theory of Numbers (Foundations of Computability). Volume III, General Semiology II. University of Maine Press, Orono, Maine.

Presburger, M.

1929. Über die Vollständigkeit eines gewissen Systems der Arithmetik ganzer Zahlen in welchem die Addition als einzige Operation hervortritt. Compte Rendus du Ier Congrès des Mathématiciens des Pays Slaves, Warszawa, 92-101.

Riemann, B.

1867. Ueber die Hypothesen, welche der Geometrie zu Grundeliegen. Abhandlungen K. Wiss, Göttingen 13.

Robbin, J. W.

1969. Mathematical Logic: A First Course. W. A. Benjamin, New York.

Robinson, A.

1966. Non-standard Analysis. North-Holland Publishing Co., Amsterdam.

Russell, B.

1919. Introduction to Mathematical Philosophy. G. Allen and Unwin Ltd., London.

Russell, B., and Whitehead, A.

1910-1913. Principia Mathematica. 2nd Edition. Cambridge University Press, Cambridge.

(1) We note that the initials are incorrectly given as W. A. Ryan in Volume 3.

Ryan, W. J.

1974. Skolem primitive recursive arithmetic $\mathscr{S}$. Master's thesis, The University of Maine at Orono.

1976. The equivalence of equational and sentential general recursive arithmetics. Journal of the London Mathematical Society (2), **14**, 463-475.

1978. Gödel's second incompleteness theorem for general recursive arithmetic. Zeitschrift für mathematische Logik und Grundlagen der Mathematik **24**, 457-459.

1979. Proof of the quadratic reciprocity law in primitive recursive arithmetic. Mathematica Scandinavica **45**, 177-197.

Schwenkel, F.

1965. Rekursive Wortfunktionen über unendlichen Alphabeten. Zeitschrift für mathematische Logik und Grundlagen der Mathematik **11**, 133-147.

Serre, J.-P.

1973. A Course in Arithmetic. Springer-Verlag, Heidelberg.

1980. Trees. Springer-Verlag, Heidelberg.

Skolem, Th.

1920. Logisch-kombinatorische Untersuchungen über die Erfüllbarkeit und Beweisbarkeit mathematischen Sätze nebst einen Theoreme über dichte Mengen. Videnskap. Skr. I, no. 4, 1-36.

1922. Einige Bemerkungen zur axiomatischen Begründung der Mengenlehre. Proc. of the 5th Scand. Math. Congress 1922, 217-232.

1930. Über einige Satzfunktionen in der Arithmetik. Videnskap. Skr. I, no. **7**, 1-28.

Suppes, P.

1960. Axiomatic Set Theory. D. Van Nostrand, Princeton.

Tarski, A.

1936. Der Wahrheitsbegriff in den formalisierten Sprachen. Studia Philosophica **1** (1936), 261-405. English translation in Logic, Semantics, Metamathematics (1956), Clarendon Press, Oxford.

Vučković, V.

1959. Partially ordered recursive arithmetics. Mathematica Scandinavica **7**, 305-320.

1962. On some possibilities in the foundations of recursive arithmetics of words. Glasnik Matematičko-Fizički I Astronomski Serija II, T. 17/No. 3-4, 145-157.

Wang, H.

1987. Reflections on Kurt Gödel. MIT Press, Cambridge, Massachusetts.

Weil, A.

1937. Sur les espaces à structure uniforme et sur la topologie général. Hermann et Cie, Paris.

1938. L'intégration dans les groupes topologiques. Hermann et Cie, Paris.

1946. Foundations of Algebraic Geometry. American Mathematical Society, New York.

1967. Basic Number Theory. Springer-Verlag, Heidelberg.

Zermelo, E.

1908. Untersuchungen über die Grundlagen der Mengenlehre I. Math. Annalen **59**, 261-281.

INDEX OF NOTATION

semicarrier space, for an atomic infinite-dimensional nonzeroed semiological space S) : §4.15

$\mathfrak{F}_{\eta SE}^{\text{scar-pre}}$, $\mathfrak{F}_{\eta SE}^{\text{scar-rec}}$, $\mathfrak{F}_{\eta SE}^{\text{scar-parec}}$ (S an atomic infinite-dimensional nonzeroed semiological space, ηSE a Peano semicarrier space) : §4.15

$\mathfrak{F}_{\eta SE}^{\text{scar-***}}$ (S an atomic infinite-dimensional nonzeroed semiological space, *** $\in$ {mon,cur,pcur}, ηSE a Peano semicarrier space) : §4.15

$\mathrm{F}_{\eta SE}^{\text{scar-}\chi\text{-***}}$ (S an atomic infinite-dimensional nonzeroed semiological space, *** $\in$ {pre,rec}, $\chi\colon \mathrm{F}_E^{***} \longrightarrow \mathrm{F}_S^{\text{total}}$, ηSE a Peano semicarrier space) : §4.15

$\mathrm{F}_{\eta SE}^{\text{scar-}\chi\text{-parec}}$ (S an atomic infinite-dimensional nonzeroed semiological space, $\chi\colon \mathrm{F}_E^{\text{parec}} \longrightarrow \mathrm{F}_S^{\text{partial}}$, ηSE a Peano semicarrier space) : §4.15

$\mathfrak{W}_{\eta SE}^{\text{scar-***}}$ (S an atomic infinite-dimensional nonzeroed semiological space, *** $\in$ {pre,rec,parec}, ηSE a Peano semicarrier space) : §4.15

$\mathrm{Wff}(\mathrm{AR}_{\eta SE}^{\text{scar-}\chi\text{-***}})$ (S an atomic infinite-dimensional nonzeroed semiological space, *** $\in$ {pre,rec}, ηSE a Peano semicarrier space, $\chi\colon \mathrm{F}_E^{***} \longrightarrow \mathrm{F}_S^{\text{total}}$) : §4.15

$\mathrm{Wff}(\mathrm{AR}_{\eta SE}^{\text{scar-}\chi\text{-parec}})$ (S an atomic infinite-dimensional nonzeroed semiological space, $\chi\colon \mathrm{F}_E^{\text{parec}} \longrightarrow \mathrm{F}_S^{\text{partial}}$, ηSE a Peano semicarrier space) : §4.15

$\mathrm{SAR}_{\eta SE}^{\text{scar-}\chi\text{-***}}$, $\mathrm{SAR}_{\eta SE}^{\chi\text{-***}}$, $\mathrm{EAR}_{\eta SE}^{\text{scar-}\chi\text{-***}}$, $\mathrm{EAR}_{\eta SE}^{\chi\text{-***}}$, $\mathrm{AR}_{\eta SE}^{\chi\text{-***}}$, $\mathrm{AR}_{\eta SE}^{***}$ (S an atomic infinite-dimensional nonzeroed semiological space, *** $\in$ {pre,rec}, $\chi\colon \mathrm{F}_E^{***} \longrightarrow \mathrm{F}_S^{\text{total}}$, ηSE a Peano semicarrier space) : §4.15

INDEX OF TERMINOLOGY